AF772648

so you want to be a HAM

by
Robert Hertzberg
K4JBI

Howard W. Sams & Co., Inc.
4300 WEST 62ND ST. INDIANAPOLIS, INDIANA 46268 USA

Copyright © 1955, 1960, 1963, 1964, 1968, 1971, 1973,
1976, and 1979 by Howard W. Sams & Co., Inc.
Indianapolis, Indiana 46268

EIGHTH EDITION
FIRST PRINTING—1979

All rights reserved. No part of this book shall be reproduced, stored in a retrieval system, or transmitted by any means, electronic, mechanical, photocopying, recording, or otherwise, without written permission from the publisher. No patent liability is assumed with respect to the use of the information contained herein. While every precaution has been taken in the preparation of this book, the publisher assumes no responsibility for errors or omissions. Neither is any liability assumed for damages resulting from the use of the information contained herein.

International Standard Book Number: 0-672-21600-0
Library of Congress Catalog Card Number: 79-63820

Printed in the United States of America.

Preface

Ham radio is one of the greatest hobbies in the world. It is so widely practiced that it is recognized and licensed by practically every country in the world. Realizing its value to national security, the United States Army, Navy, and Air Force sponsor ham-radio organizations. It is encouraged by the electronics industry because of its importance to technical progress. Municipal officials everywhere have praised the contributions of ham radio to public service and disaster aid.

Thousands of new enthusiasts are attracted to ham radio each year. This book is intended to get beginners of all ages off to the right start in this fascinating hobby; it also serves as an invaluable reference for the old timer who has been "pounding brass" for years.

The entire field of amateur radio is encompassed in this volume —from the various types of licenses, and the requirements and procedures for obtaining them, to painless methods for learning the code. This book covers the value of kit building and the practical knowledge and experience gained in assembly. Completely updated, other chapters are devoted to receivers, transmitters, mobile equipment, test equipment, antenna systems, and 2-meter fm. In addition, such diversified subjects as operating procedures, the organization of amateur radio, and where and how to purchase equipment are included. The chapter on electronics as a career will be of particular value to young men.

Robert Hertzberg K4JBI

Contents

1

The Amateur Radio Game

It is late in the afternoon of a wintry day and you're killing time before supper. All you can find on tv is an old Western in which the "good guy" manages to get nine or ten shots out of a six-shooter without reloading. You decide to warm up your CB rig, which you haven't used much recently because a lot of illegal "rock-crushers" have been souring the local channels. You punch up Channel 11, and a familiar voice punctures the interference. It belongs to "Hairy Harry," who runs an appliance repair service. He's in his truck, getting a message from his wife back at the shop. After he closes with 10-10, you press your own mike button and give him a shout.

"Hairy Harry, this is Big Joe. Long time no see. Got any new gear lately?"

This turns out to be a loaded question.

"Man, I should say I have," he comes back. "Only it's not 5-watt CB, but 150-watt ham equipment and it's all legit. I have a ham license and I'm having a ball!"

"Sounds great," you say, not very enthusiastically. "Talk to anybody yet with it?"

"Talk to anybody? You gotta be kidding," he snaps back. "Right now I'm using only a hunk of hookup wire for an antenna, but the first week I had it on the roof I chewed the fat with hams in England and Germany."

Your curiosity is aroused.

"You've been holding out on me, buddy boy. I didn't know you spoke German."

"No sweat," says Harry. "Those guys talk better than I do. Funny thing is that I had trouble understanding the English operators. They sounded like they had loose teeth."

"You gonna sell your CB stuff?" You pose the question carefully, thinking you might buy some of it at bargain prices.

"Heck, no. It's fine for my small business, but the ham station is for fun. Hey, I'm almost home. Will buzz you on the land line after chow."

"10-4, Harry. See you later."

You've experienced occasional 1000-mile "skip" yourself on CB, but England and Germany are something else. Was Harry only joking?

He does phone back, and invites you to come over and see his new toy. Some toy! You take one look and immediately your fingers start to itch; on the front panel there are 19 knobs and 14 other controls. You can figure that MIC means microphone, but what about VOX, MONI, DH, FSK, RIT, etc?

"This is over my head," you admit, "but I do recognize the switch marked POWER ON. May I push it?"

"Sure. See what happens."

Instantly, the figures 14.278.2 appear in bright green in a little window over the big center tuning knob and the loudspeaker spews out a jumble of strong signals.

"You are tuned to 14.278 megahertz plus 200 hertz," Harry explains. "This is in the phone section of the 20-meter band, which is great for DX. The calibration coincides to the last hertz with the signals of WWV, the station of the Bureau of Standards. Are you still with me on all this?"

"He better not be," interposes a new voice. Harry's wife is in the doorway. "His wife called to remind him that he goes on the early work shift tomorrow."

You'd like to stay, because suddenly you want to ask a lot of questions about ham radio. Welcome to the club!

WHAT'S A "HAM"?

One of the first things you learn is that amateur radio operators rarely call themselves that; they are *hams*. In other fields of activity, particularly the theater, calling a man a "ham" is likely to get you a swift right to the jaw, but in radio circles the term is one of distinction. Practicing this great hobby are doctors, lawyers, and Indian chiefs; generals, admirals, and lowly seamen; corporation presidents, bus drivers, and grocery clerks; retired octogenarians and boys whose voices haven't changed yet; grandmothers and teenage girls; even the king of an Asiatic country and a one-time candidate for the presidency of the United States. All these amateur radio operators are proud to be "hams." The origin of the word is a complete mystery, even to hams who have been hams for fifty years and more.

Amateur radio is almost as old as radio itself. Intrigued by the possibilities of communication through space, electrical experimenters at the turn of the last century emulated Marconi by building crude but successful receivers and transmitters. By 1910 there were several hundred such "wireless amateurs" in the United States alone, and others in Europe.

The technical contributions to the art made by amateurs, and their right to maintain private stations for experimental and noncommercial purposes, were recognized and affirmed in the first Federal legislation regulating radio communication passed by Congress in 1912. A licensing procedure was set up, and certain restrictions were imposed on amateur operation to prevent interference with vital ship-to-shore communication. Power was limited to one kilowatt, and the maximum wavelength to 200 meters. (This is now the bottom edge of the regular broadcast band.) The wavelengths in general use during this period ran from 600 meters and upward. Many technicians thought that anything as short as 200 meters was useless, and that the poor hams would soon die off because they'd be unable to communicate with each other. The hams promptly proceeded to prove otherwise, and today receive full credit for the extraordinary development of short-wave radio as we now know it.

FCC LICENSES

Radio laws were first administered by the Department of Commerce, then by the Federal Radio Commission, and now by the all-encompassing Federal Communications Commission (FCC). The government attitude toward hams has consistently been paternalistic and protective; hence, ham radio as a hobby has had a steady and ever-increasing growth.

No license of any kind is needed for short-wave receivers, any more so than for broadcast receivers or television sets. However, you must have an FCC Amateur Radio License to operate transmitting equipment for communicating with other amateurs, regardless of how small or low-powered it is. There's no limit on the number of transmitters you can own and use under a single license—you can fill a whole basement with them, put one in your car, boat or airplane, carry one on your back, and have still others at a camp or cottage.

All ham licenses are issued without cost.

Eligibility for Licenses

Any citizen or national of the United States is eligible for a ham license, or "ticket." Get this: *any* citizen, man or boy, woman or girl, healthy or disabled. Age is no bar. The only requirements are the ability to send and receive the International Morse Code, a knowledge of a few technical fundamentals, and a familiarity with FCC regulations. In 1948, two brothers aged eleven and nine, respectively, both passed the FCC examination. They were Ken William Lattig, W9FZE, and Lowell Kay Lattig, W9FZJ. Probably the outstanding junior of all time was Jean Hudson, who in 1934 or thereabouts flabbergasted FCC inspectors by copying 20 words per minute. She was then just eight years old! At the time Jean and the Lattig boys got their tickets the examination was much more difficult than it is now.

Michael Homenick, WA2PAO, Flushing, N.Y., was scarcely 15 when he obtained his ham license. Using low-power equipment, most of it home-assembled, he has communicated by means of the International Morse Code with stations all over the world, and he can prove it by the acknowledgement cards he has received from them. His entire "radio shack" is a basement closet only four feet wide.

The inauguration of the relatively easier "novice" and "technician class" licenses enabled many would-be hams to qualify for licenses with less preparation than previously required. The FCC has records of three seven-year-olds who made the novice license

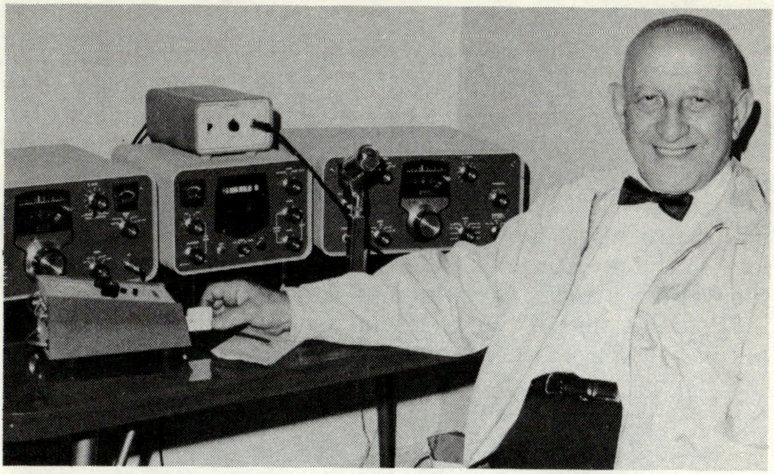

At the other end of the age scale is Frank Frimerman, W2FZ. When this picture was taken in 1970 he was hale and hearty at 75 and has been an active ham for more than half a century. In fact, in the 1920s, before the advent of commercial broadcasting, he was on the air with weekly phonograph programs, much to the delight of owners of early crystal detector receivers in New York City.

on the first try. One was a girl. Of the boys, one found blindness no handicap in achieving his aim.

The enthusiasm that fired these youngsters was shared by Dr. J. Van Becelaere, a retired physician of San Diego, Cal. He got his ticket when he was a mere 86 years old!

In many households ham radio is a family affair. The contagion starts with a parent or child and works out in either direction. Families in which father, mother, and several children are all licensed amateurs are now common and no longer attract attention. There are numerous cases of "radio romances" that started with conversations over the air and culminated in happy marriages. Since the children of these couples learn to recognize a microphone about the same time they learn to feed themselves, it's no wonder that they become hams at a tender age.

By the middle of 1979 there were more than three hundred thousand amateur licenses in effect in the United States and its possessions and territories. There are probably four or five times that number of ex-hams who have let their tickets lapse over the years but who retain their interest in the game, usually by continuing to use short-wave receivers. In other countries of the world there are about 50,000 licensed operators. The majority of them speak English, which is now indubitably the international language.

Standard Types of Licenses

There are five grades of ham licenses: *novice, technician, general, advanced, and amateur extra.* As their names imply, they require different degrees of technical know-how; also, they offer progressively better operating privileges on the air. The licensing procedure is described in detail in Chapter 5.

With your license, the FCC gives you a "call sign" consisting of one or two letters, a number from 1 through Ø, and two or three letters. This is your only radio identity. Hams never, but never, use weird concocted names of the CB variety.

To appreciate the differences among the various licenses, it is necessary to have a picture of the divisions of the seven most pop-

Living in a crowded New York apartment is no deterrent to a determined ham. Here, a board across a windowsill holds a Collins receiver and transmitter without blocking the view of the Hudson River and of the New Jersey cliffs across from Manhattan. An automobile type antenna protruding from under another window permits plenty of DX contacts.

Table 1-1. Frequency Division in Ham Bands

	Frequency (MHz)	
Band	*Entire Band*	*Phone Section*
80 Meters	3.500-4.000	3.775-4.000
40 Meters	7.000-7.300	7.150-7.300
20 Meters	14.000-14.350	14.200-14.350
15 Meters	21.000-21.450	21.250-21.450
10 Meters	28.000-29.700	28.500-29.700
6 Meters	50.000-54.000	50.100-54.000
2 Meters	144.000-148.000	144.100-148.000

ular ham bands (Table 1-1). The latter are designated roughly in terms of wavelength as *80 meters, 40 meters,* etc., and more accurately in terms of frequency. The abbreviation *MHz* stands for *megahertz*. The prefix *mega,* of course, means *million*. The unit *hertz* was adopted by international technical societies in the 1960s to honor Heinrich Hertz, a German scientist whose experiments in the 1880s led to the practical development of the "wireless." The term *hertz* has exactly the same meaning as *cycles per second,* and takes various common prefixes such as *kilo, mega, giga,* etc.

This lady isn't posed in the picture just to give it human interest. She makes good use of this almost bewildering array of receivers and transmitters, acquired by her and her husband over a period of years. She is Vera K. Woods, W7TGG, Brady, Montana, and her "old man" is Ray, W7SFK. They must be the local power company's best customer.

Technically, radiotelegraphy is permitted in all portions of all bands, while radiotelephony is permitted only in the assigned phone sections as shown in Table 1-1. However, it is very unusual to hear anything but voice on the phone frequencies.

The novice license, as you might expect, is the easiest to obtain. You must be able to send and receive code at the rate of five words per minute and pass a simple examination covering basic radio theory and operating procedures. Considering the ages of the boys and girls who sail through the tests without strain, the novice ticket is, literally, kindergarten stuff. Operating privileges are a bit limited, but at least you can get on the air quickly. You may use only radiotelegraphy, with a maximum of 250 watts of transmitter power, on narrow segments of the 10-, 15-, 40- and 80-meter bands. *You are not permitted to use voice transmission at all.*

Modest, but busy, ham station in the corner of a room. On table, left to right, Electro-Voice 664 microphone, Heathkit SB-102 five-band transceiver, Italian-made Geloso general-coverage receiver. Accessories on shelf: three-inch oscilloscope, 24-hour clock on antenna tuning indicator, speaker, and callbooks. Operator is using earphones to avoid bothering tv listeners in the same room.

The technician license also requires the five-word code test, but the written test calls for more technical knowledge. In addition to the novice privileges, you may operate telegraph and voice with as much power as 1000 watts on all amateur frequencies from 50 MHz and upward. The important part of this provision is that it includes the highly popular 2-meter fm band. (See Chapter 12.)

To qualify for a general class "ticket" you must get up to 13 words per minute with the code and pass a written test dealing with general amateur practices and the FCC regulations. All of some bands and major parts of the others are open to holders of this ticket. The exclusions comprise certain segments reserved for advanced and amateur extra operators.

The advanced license, the next higher grade, consists of the 13 word-per-minute code test and a slightly more extensive written exam. The highest license, the amateur extra, really should be called the *expert* because it calls for code at 20 words per minute and a really tough technical and operating exam. The only way to build up code speed is by practice, so most holders of the extra ticket are hams with a lot of experience on the air.

These restrictions and allotments aren't as bad as they sound. Actually, the bands are very wide in terms of tuning space on the

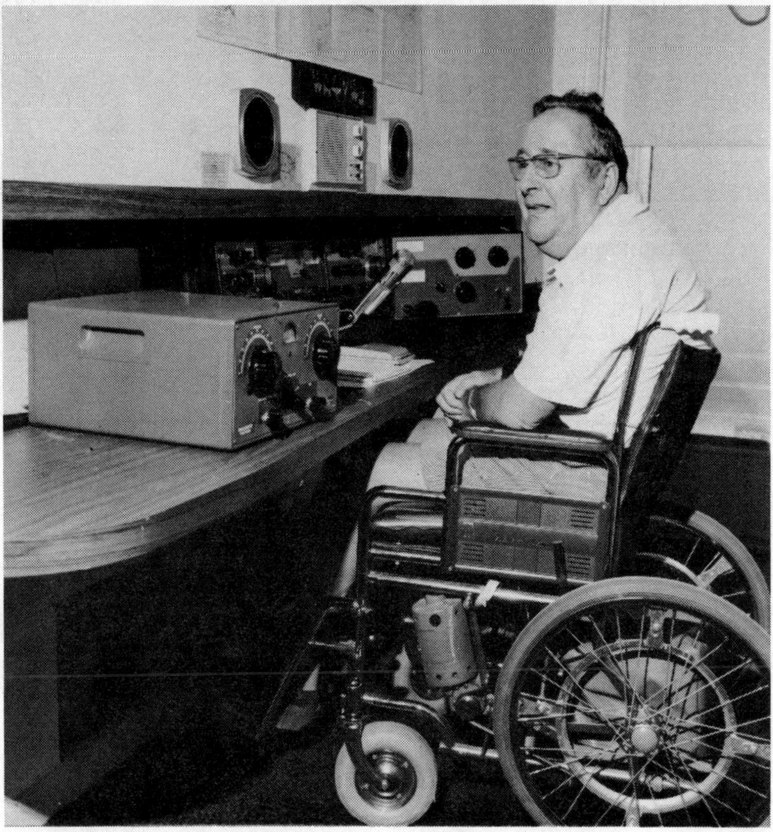

Look closely at this picture. Would you believe that the occupant of the motor-driven chair, a disabled victim of multiple sclerosis, has been called a life saver by grateful people who have never met him? He is Francis Healey of Oakland Park, Florida, better known to fellow hams on several continents as WA4VWJ. On one occasion he was able to organize a rescue party to an isolated section of Honduras; on another occasion, he helped to save a boatload of passengers on a stricken vessel off Guatemala. He has spent as much as eleven hours a day at his station, which was assembled for him by friends. What a wonderful activity ham radio is for this ex-Marine and for many other men and women in his position!

dials of transmitters and receivers, and some sections of the bands are relatively uninhabited.

All five ham licenses are good for five years and can be renewed successively for further five-year periods. There is no government charge for the examinations or the licenses.

This division of the ham frequencies does not mean that holders of one class of license must communicate only with other holders of the same ticket. All extra, advanced, and general operators are free to move into the novice and technician bands and to hobnob with the licensees here, giving them code practice, technical assistance, operating hints, etc. However, novices and technicians may not go into the segments reserved for the other classes. This sounds like discrimination, and it surely is, but it only encourages beginners to elevate themselves quickly into the higher grades.

The regulations governing amateur operations are subject to frequent change. For current information consult the magazines *QST*, *CQ*, *73*, and *Ham Radio*, which are available from most ham radio dealers.

CODE VS. PHONE

Some newcomers to the game are bothered by the necessity of learning the code. "I intend to use only voice operation. Why should I mess with dits and dahs?" is a common complaint. It is only after a beginner masters the code that he begins to appreciate its usefulness, and also to understand why the FCC and the great majority of the hams themselves insist on keeping it as a basic requirement for a license.

Communication by dits and dahs is usually called "CW" operation. (The letters stand for "continuous wave.") The dits and dahs are merely short and long spurts of these waves, made by the action of a hand-operated telegraph key. In radiotelephone operation, usually referred to simply as "phone," the same continuous waves are modulated to conform to the voice of the operator as he impresses it on a microphone, or "mike." A CW signal occupies a very sharp, narrow slice of space on the air, while a phone signal needs a much wider slot. You might liken the former to a motorcycle, which can weave in and out of traffic readily; and the latter to a passenger car, which is much less maneuverable.

CW signals convey intelligence only by their length, not by their musical quality or tone. Any short sound or noise is a dit (sometimes called a dot), a slightly longer one a dah (sometimes called a dash). CW signals can be weak, distorted, hacked up by interference, almost obliterated, yet a skilled operator can usually catch enough letters or whole words to form an understandable message. Many experienced hams can even "read" two simultaneous signals if they are of slightly different pitch!

Phone signals, on the other hand, are much more vulnerable to the effects of noise and interference. Since voice intelligence is conveyed by changes and shading in enunciation, inflection, etc., it is not surprising that the usefulness of received signals drops very rapidly if they are broken up by other modulated signals on

Two duplicate operating positions, five separate kilowatt amplifiers for the bands 10 through 80 meters, five separate antennas, and a generally elegant atmosphere are some of the features of station W2AB, assembled over a period of years by Lawrence LeKashman, Long Island, N.Y. A noted winner of operating contests, Larry can work on two bands at the same time by means of split headphones! Head of an electronics firm, he is one of many active hams who work professionally in electronics and enjoy radio as a hobby.

closely adjacent frequencies. Phone signals can be very loud and still have no more meaning than the sounds in the testing room of a saxophone factory.

Transmitter power, operating frequency, atmospheric conditions, etc., being equal, CW and phone signals will travel equally as far. However, because CW signals can be separated more easily in a receiver and are understandable even if weak and distorted, they provide useful communication in circumstances that render phone signals useless. From the practical standpoint, this means that you can communicate with CW over much greater distances with much less trouble than with phone. Putting it another way, you need much more power on voice to achieve the same relative degree of reliability as on CW.

Many, many years of experience bear this out. With simple, low-powered CW transmitters one-third the size of a standard short-wave receiver, hams have no trouble at all working around the globe. Some years ago an amateur in the Bronx, in New York City, was the most reliable contact with an expedition in the wilds of South America. Day after day he received dispatches from the explorers and sent back news and other information, all on CW.

What's a tv camera doing in this beautiful ham shack of John A. Smetona, K3SLJ, of Pottsville, Pennsylvania? Transmitting slow-scan pictures of John to other hams, that's what! Note monitor receiver under the clock. This television equipment, designed especially for amateur purposes, is made by Robot Research, Inc., San Diego, California.

He tried to keep the details of his transmitter a secret, because he felt that no one would believe him. The rig used one, solitary receiving tube of a type then called "peanut" (today we'd call it "miniature"), energized by three 45-volt "B" batteries and drawing the grand power of exactly one watt!

You may learn the code only because of the FCC requirements, but you will soon become quite proud of your ability to sling the dits and dahs. You will probably discover, as many hams do eventually, that it is much more challenging and sporting to work a New Zealander with one to ten watts of CW than to contact a North African with 100 or 1000 watts of phone. It's like in golf. Which is more satisfying . . . dropping a ball into a tiny cup at 200 yards with one stroke, or sinking an easy putt at two feet?

AMATEUR TRANSMISSIONS

Hams do not "broadcast." They are specifically prohibited from transmitting music and material of general listening interest. Hams "send" or "transmit" only to each other, under ordinary circumstances; in amateur lingo this is called "working." The usual forms of communication are phone and CW, with radioteletype (RTTY) a popular third. The ultimate in person-to-person contact on the air is unquestionably ham television (yes, *television!*) which is permitted on frequencies of 420 megahertz and upward. This is not full-scale tv such as you see in color in the home, but is a relatively simplied version known as "slow scan." The equipment is rather expensive, but this doesn't stop an increasing number of hams from experimenting with it. Several manufacturers are offering cameras, monitors, receivers, etc.

HAM RADIO AND CB

There is absolutely no connection of any kind, legally or technically, between amateur radio and Citizens-Band operation. Originally, CB was primarily intended for communications between units of one owner; for example, a central control station in the office of a delivery service and individual trucks on the road. In this sense CB was intended to serve small business firms. Recent changes in the FCC Rules and Regulations pertaining to CB allow the CB operators to communicate with other CB stations for almost any legal purpose. However, the input power to CB transmitters is still limited to 5 watts and communications are limited to distances of less than 150 miles. In practice, CB communications are generally limited to 10 miles or less. A CB license does not require a technical examination of any kind, whereas all grades of the ham ticket involve a two-part test. A ham ticket is not valid for CB purposes and a CB ticket is not valid for ham purposes. However, there is nothing to prevent you from holding both types of licenses and to use them independently of each other.

THE HAM'S POSITION IN THE RADIO SPECTRUM

The frequencies allotted to amateur radio are only small fillers in an enormous electronic sandwich whose bottom slice is the 10-kilohertz lower limit of the long-range navigation stations and whose top is the 275-gigahertz edge of satellite assignments. In the United States alone, there are probably three million or more licensed stations of various kinds, and there are thousands more in other countries. Therefore, a prime responsibility of hams everywhere is to stay on their frequencies with properly adjusted equipment that does not cause interference to any other service. Happily, hams over the years have earned a remarkable reputa-

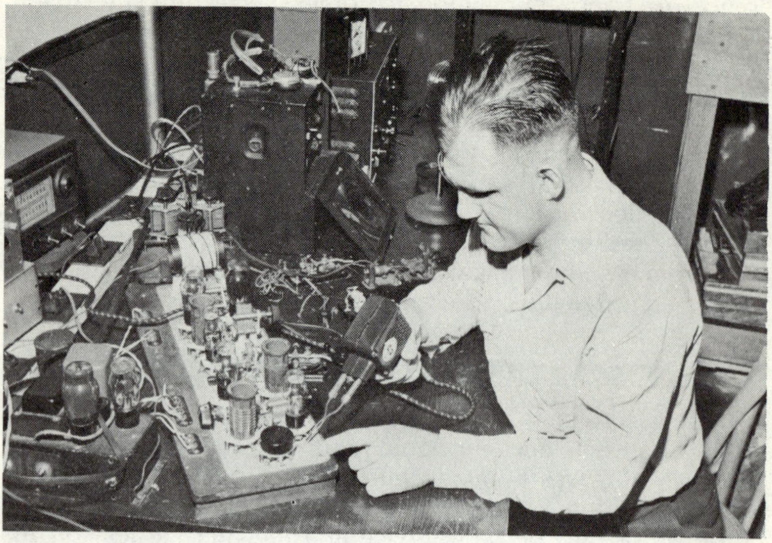

There is nothing unusual about this picture of a ham wiring up a "breadboard" except for one thing: He happens to be totally blind! Blindness did not stop Robert W. Gunderson. W2JIO, from becoming active as a teacher of radio as well as the designer of a line of electronic gear that is "read" by the fingers of of sightless persons.

tion for technical competence and their ability to police their own bands; in no small way, this accounts for the privileges they enjoy.

Actual revocations of licenses for purposeful infractions of FCC regulations are extremely rare, and even temporary suspensions for minor incidents number hardly a dozen a year. This is indeed a good record when you consider the number of amateur stations on the air.

FCC REGULATIONS

The ham's code of behavior is set forth in the *Rules and Regulations of the Federal Communications Commission* Part 97, "Amateur Radio Service." This was formerly a part of Volume VI, which is no longer issued. Part 97 can be purchased by mail from the Superintendent of Documents, U.S. Government Printing Office, Washington, D.C. 20402. The Stock No. is 004-000-00357-8, and the price is $1.40. Send a check or money order, not cash. This booklet is not available from the FCC itself.

A few of the rules of immediate interest to newcomers are reprinted here, with comments the author thinks might be helpful.

SUBPART A—GENERAL

§ 97.1 Basic and purpose:

The rules and regulations in this part are designed to provide an amateur radio service having fundamental purpose as expressed in the following principles:

(a) Recognition and enhancement of the value of the amateur service to the public as a voluntary noncommercial communication service, particularly with respect to providing emergency communications.

(b) Continuation and extension of the amateur's proven ability to contribute to the advancement of the radio art.

(c) Encouragement and improvement of the amateur radio service through rules which provide for advancing skills in both the communication and technical phases of the art.

(d) Expansion of the existing reservoir within the amateur radio service of trained operators, technicians, and electronics experts.

(e) Continuation and extension of the amateur's unique ability to enhance international good will.

§ 97.3 Definitions.

(a) *Amateur radio service.* A radio communication service of self-training, intercommunication, and technical investigation carried on by amateur radio operators.

(c) *Amateur radio operator* means a person holding a valid license to operate an amateur radio station licensed by the Federal Communications Commission.

§ 97.27 Mail examinations for applicants unable to travel.

The Commission may permit the examinations for an Amateur Extra, Advanced, General, or Technician Class license to be administered at a location other than a Commission examination point by an examiner chosen by the Commission when it is shown by physician's certification that the applicant is unable to appear at a regular Commission examination point because of a protracted disability preventing travel.

§ 97.28 Manner of conducting examinations.

(d) All written portions of the examinations for amateur operator privileges shall be completed by the applicant in legible handwriting or hand printing. Whenever the applicant's signature is required, his normal signature shall be used. Applicants unable to comply with these requirements, because of physical disability, may dictate their answers to the examination questions and the receiving code test. If the examination or any part thereof is dictated, the examiner shall certify the nature of the applicant's disability and the name and address of the person(s) taking and transcribing the applicant's dictation.

§ 97.37 General eligibility for station license.

An amateur radio station license will be issued only to a licensed amateur radio operator, except that a military recreation station license may also be issued to an individual not licensed as an amateur radio operator (other than a representative of a foreign government), who is in charge of a proposed military recreation station not operated by the U.S. Government but which is to be located in approved public quarters.

Comment

The last portion of paragraph 97.37 takes care of Naval Reserve, Military Affiliate Radio System and ham club stations at various military establishments. The licensee in these cases is usually the commanding officer of the unit or activity through which the station is administered.

§ 97.39. Eligibility of corporations or organizations to hold station license.

An amateur station license will not be issued to a school, company, corporation, association, or other organization, except that in the case of a *bona fide* amateur radio organization or society meeting the criteria set forth in Section 97.3, a station license may be issued to a licensed amateur operator, other than the holder of a Novice Class license, as trustee for such society.

Comment

The first part of paragraph 97.39 is in line with the idea that amateur radio is strictly amateur. The second part covers civilian radio clubs. In connection with the first part, it should be pointed out that a collection of equipment in the store or showroom of a radio dealer, jobber, or other commercial sales agency does not constitute an amateur radio station, and should not be put on the air for so-called demonstration purposes either by licensed employees of the firm or by visiting hams who might be prospective customers. This is illegal operation from several standpoints.

§ 97.43 Location of station.

Every amateur station must have at least one land location, the address of which appears on the station license and at least one control point.

§ 97.63 Selection and use of frequencies.

(a) An amateur station may transmit on any frequency within any authorized frequency band.

(b) Sideband frequencies resulting from keying or modulating a carrier wave shall be confined within the authorized amateur band.

§ 97.79 Control operator requirements.

(a) The licensee of an amateur station shall be responsible for its proper operation.

(b) Every amateur radio station, when in operation, shall have a control operator at an authorized control point. The control operator shall be on duty, except where the station is operated under automatic control. The control operator may be the station licensee, if a licensed amateur radio operator, or may be another amateur radio operator with the required class of license and designated by the station licensee. The control operator shall also be responsible, together with the station licensee, for the proper operation of the station.

(c) An amateur station may only be operated in the manner and to the extent permitted by the operator privileges authorized for the class of license held by the control operator, but may exceed those of the station licensee provided proper station identification procedures are performed.

(d) The licensee of an amateur radio station may permit any third party to participate in amateur radio communications from his station, provided that a control operator is present and continuously monitors and supervises the radio communication to insure compliance with the rules.

Comment

The last part of this paragraph contains a wide open loophole in the form of the words ". . . . any third party . . ." This means that a visitor doesn't need a license of any kind to talk over a phone station, providing the owner is present and handles the equipment as prescribed. The main purpose of the provision is to permit hams to furnish free person-to-person voice communication for people separated great distances (see Chapter 9). It also gives them the chance to put visitors on the air briefly for what thrills they get out of the experience.

§ 97.82 Availability of operator license.

The original operator license of each operator shall be kept in the personal possession of the operator while operating an amateur station. When operating an amateur station at a fixed location, however, the license may be posted in a conspicuous place in the room occupied by the operator. The license shall be available for inspection by any authorized Government official whenever the operator is operating an amateur station and at other times upon request made by an authorized representative of the Commission, except when such license has been filed with application for modification or renewal thereof, or has been mutilated, lost or destroyed, and request has been made for a duplicated license in accordance with § 97.57. No recognition shall be accorded to any photocopy of an operator license; however, nothing in this section shall be construed to prohibit the photocopying for other purposes of any amateur radio operator license.

§ 97.83 Availability of station license.

The original license of each amateur station or photocopy thereof shall be posted in a conspicuous place in the room occupied by the licensed operator while the station is being operated at a fixed location or shall be kept in his personal possession. When the station is operated at other than a fixed location, the original station license or a photocopy thereof shall be kept in the personal possession of the station licensee (or a licensed representative) who shall be present at the station while it is being operated as a portable or mobile station. The original station license shall be available for inspection by any authorized Government official at all times while the station is being operated and at other times upon request made by an authorized representative of the Commission, except when such license has been filed with application for modification or renewal thereof, or has been mutilated, lost, or destroyed, and application has been made for a duplicate license in accordance with § 97.57.

Comment

From reading paragraphs 97.82 and 97.83, you might get the idea that the operator license and the station license are two separate documents. Actually, they are a single piece of paper 7⅜ inches long by 2¼ inches wide, folded to about the size of a credit card. One side is marked "Amateur Radio License" and includes the name and address of the licensee; the other side has the assigned call sign, operator privileges, and related data.

These paragraphs are a bit contradictory. The first reads, "no recognition shall be accorded to any photocopy of an operator license," but the second paragraph says, "The original license of each amateur station or a photocopy thereof shall be posted. . . ." Your ticket is valuable; protect it as you do other personal records.

§ 97.89 Points of communications.

(a) Amateur stations may communicate with:

(1) Other amateur stations except those prohibited by Appendix 2.

(2) Stations in other services licensed by the Commission and with U.S. Government stations for civil defense purposes in accordance with Subpart F of this part, and on a temporary basis, for test purposes.

(3) Any station which is authorized by the Commission to communicate with amateur stations.

(b) Amateur stations may be used for transmitting signals, or communications, or energy, to receiving apparatus for the measurement of emissions, temporary observation of transmission phenomena, radio control of remote objects, and similar experimental purposes and for purposes set forth in § 97.91.

§ 97.91 One-way communications.

In addition to the experimental one-way transmissions permitted by § 97.89, the following kinds of one-way communications, addressed to amateur stations, are authorized and will not be construed as broadcasting: (a) Emergency communications, including bona-fide emergency drill practice transmissions; (b) Information bulletins consisting solely of subject matter having direct interest to the amateur radio service as such; (c) Round-table discussions or net-type operations where more than two amateur stations are in communication, each station taking a turn at transmitting to other station(s) of the group; and (d) Code practice transmission intended for persons learning or improving proficiency in the International Morse Code.

§ 97.101 Mobile stations aboard ships or aircraft.

In addition to complying with all other applicable rules, an amateur mobile station operated on board a ship or aircraft must comply with all of the following special conditions: (a) The installation and operation of the amateur mobile station shall be approved by the master of the ship or captain of the aircraft; (b) The amateur mobile station shall be separate from and independent of all other radio equipment, if any, installed on the ship or aircraft; (c) The electrical installation of the amateur mobile station shall be in accord with the rules applicable to ships or aircraft as promulgated by the appropriate government agency; (d) The operation of the amateur mobile station shall not interfere with the efficient operation of any other radio equipment on board the ship or aircraft; and (e) The amateur mobile station and its associated equipment, either in itself or in its method of operation, shall not constitute a hazard to the safety of life or property.

§ 97.112 No remuneration for use of station.

(a) An amateur station shall not be used to transmit or receive messages for hire, nor for communication for material compensation, direct or indirect, paid or promised.

§ 97.113 Broadcasting prohibited.

Subject to the provisions of § 97.91, an amateur station shall not be used to engage in any form of broadcasting, that is, the dissemination of radio communications intended to be received by the public directly or by the intermediary of relay stations, nor for the retransmission by automatic means of programs or signals emanating from any class of station other than amateur. The foregoing provision shall not be construed to prohibit amateur operators from giving their consent to the rebroadcast by broadcast stations of the transmissions of their amateur stations, provided, that the transmissions of the amateur stations shall not contain any direct or indirect reference to the rebroadcast.

§ 97.115 Music prohibited.

The transmission of music by an amateur station is forbidden.

§ 97.117 Codes and ciphers prohibited.

The transmission by radio of messages in codes or ciphers in domestic and international communications to or between amateur stations is prohibited. All communications regardless of type of emission employed shall be in plain language except that generally recognized abbreviations established by regulation or custom and usage are permissible as are any other abbreviations or signals where the intent is not to obscure the meaning but only to facilitate communications.

2

Conquering the Code

The radio code is known officially as the "International Morse Code." The letters of the alphabet, the ten numbers, the punctuation marks and certain special characters are represented by combinations of short and long sounds. These are often referred to as "dots'" and "dashes" because that's a convenient way of showing them in printed charts. A more accurate approach is to call the sounds "dits" (usually referred to as *di* except when it is the last element within a character) and "dahs." Ideally, the dah sound is supposed to be three times longer in duration than the very short dit sound. The spacing or silent period between dits and dahs of the same character is equivalent to the duration of one dit; between letters three dit spaces; and between words five dit spaces. Actually, it makes little difference how short or long the sounds are, providing only that the dahs are noticeably longer than the dits. However, the spacing is important, and you'll understand why when you look over Chart 2-1.

Note that a single dit represents the letter *E,* a single dah is *T,* and a dit followed by a dah is *A.* The word eat would thus be sent:

dit di-dah dah

If the sending is sloppy and the first three signals are run together, the transmission becomes:

di-di-dah dah

The first character is the letter *U,* so the word comes out *ut* if you can call that a word.

If the last three signals are run together, the transmission sounds like:

dit di-dah-dah

The second letter is now *W* and the word is *ew,* again, not even remotely connected with *eat.*

Chart 2-1. The International Morse Code

ALPHABETICALLY

A—di-dah
B—dah-di-di-dit
C—dah-di-dah-dit
D—dah-di-dit
E—dit
F—di-di-dah-dit
G—dah-dah-dit
H—di-di-di-dit
I—di-dit
J—di-dah-dah-dah
K—dah-di-dah
L—di-dah-di-dit
M—dah-dah
N—dah-dit
O—dah-dah-dah
P—di-dah-dah-dit
Q—dah-dah-di-dah
R—di-dah-dit
S—di-di-dit
T—dah
U—di-di-dah
V—di-di-di-dah
W—di-dah-dah
X—dah-di-di-dah
Y—dah-di-dah-dah
Z—dah-dah-di-dit

BY GROUPS

Group 1
E—dit
I—di-dit
S—di-di-dit
H—di-di-di-dit
5—di-di-di-di-dit
T—dah
M—dah-dah
O—dah-dah-dah

Group 2
A—di-dah
W—di-dah-dah
J—di-dah-dah-dah
N—dah-dit
D—dah-di-dit
B—dah-di-di-dit

Group 3
R—di-dah-dit
F—di-di-dah-dit
L—di-dah-di-dit
U—di-di-dah
V—di-di-di-dah

Group 4
K—dah-di-dah
X—dah-di-di-dah
C—dah-di-dah-dit
Y—dah-di-dah-dah
Q—dah-dah-di-dah
G—dah-dah-dit
Z—dah-dah-di-dit
P—di-dah-dah-dit

NUMERALS

1—di-dah-dah-dah-dah
2—di-di-dah-dah-dah
3—di-di-di-dah-dah
4—di-di-di-di-dah
5—di-di-di-di-dit
6—dah-di-di-di-dit
7—dah-dah-di-di-dit
8—dah-dah-dah-di-dit
9—dah-dah-dah-dah-dit
0—dah-dah-dah-dah-dah

PUNCTUATION

Period—di-dah-di-dah-di-dah
Comma—dah-dah-di-di-dah-dah
Semicolon—dah-di-dah-di-dah-dit
Colon—dah-dah-dah-di-di-dit
Question Mark—di-di-dah-dah-di-dit
Double Dash—dah-di-di-di-dah
Fraction Bar—dah-di-di-dah-dit
Hyphen—dah-di-di-di-di-dah
Parenthesis—dah-di-dah-dah-di-dah

SPECIAL CHARACTERS

Wait—di-dah-di-di-dit
Invitation to Transmit—dah-di-dah
End of Message—di-dah-di-dah-dit
End of Work—di-di-di-dah-di-dah
Error—di-di-di-di-di-di-di-dit

THE WIRE TELEGRAPH CODE

International Morse is often called "Continental Morse" or simply "Continental," because it was used on European wire telegraph systems for years before the invention of "wireless." Don't confuse it with "American Morse," which was named for Samuel F. B. Morse, the inventor who made wire line telegraphy practicable. In both codes the characters consist of dits alone, dahs alone, and combinations of dits and dahs. Fifteen characters of the alphabet and one number use the same combinations in both. In International Morse, the spacing within individual characters is uniform. In American Morse, however, variations in the spacing within some characters determine the identity of the letter. The best example is the combination of three dits. Sent straight, with the spacing between dits equal in duration to a single dit, it is the letter *S* in both codes:

dit (one space) dit (one space) dit

In American Morse a combination of three dits has two additional meanings, depending on where a double-dit spacing is used. Thus:

dit (one space) dit (two spaces) dit—the letter C,
dit (two spaces) dit (one space) dit—the letter *R*.

In both codes the letter *I* is di-dit, but in American Morse, when the two dits are separated by a double space, they become the letter *O*!

The opportunities for confusion are obviously great. You need a very sharp ear to distinguish between dit dit dit for *S* and di-dit dit for *C;* in fact, you're quite likely to write down *ie* for the latter.

In wire line telegraphy, all the sounds emitted by the receiving instrument are actually individual clicks produced by a spring-loaded brass lever striking against a U-shaped brass anvil. Appropriately enough, this instrument is called a "sounder." When a transmitting operator taps his key once to make the letter *E,* the sounder makes one click going down and another when it springs back. When he holds his key down to make the single dah of the letter *T,* the sounder again makes one click going down and another going up, and the dah is indicated by the length of the silence between the two clicks. Tricky? It sure is. In radio work, the dits and dahs are all positive buzzes or whistling sounds; therefore they are easier to recognize and memorize.

During the period between the Civil War and World War I, clicking sounders in railroad stations and telegraph offces represented the major means of electrical communication. This service soon gave way to the telephone and the teletypewriter. However, in recent years there has been a surprising revival of interest in American Morse, not among old-timers but rather among younger CW operators who wouldn't recognize a sounder if they saw one. There's even a national Morse Club, devoted to the use of the venerable American Morse code on the ham bands.

CODE PRACTICE SETS

"What's the best way to start on the code?" you're probably asking by now. The very first thing you should do is to buy or make some sort of code-practice set. This is generally a simple mechanical or electronic device that generates whistling or buzzing sounds that resemble real CW signals. The set illustrated in the accompanying photograph consists of a standard radio "key," a common house buzzer, and two flashlight cells. For a baseboard use any clean piece of wood about 5 by 6 inches. Mount the key and the buzzer with short wood screws. Place the batteries so that the center contact of one is adjacent to the plain bottom of the

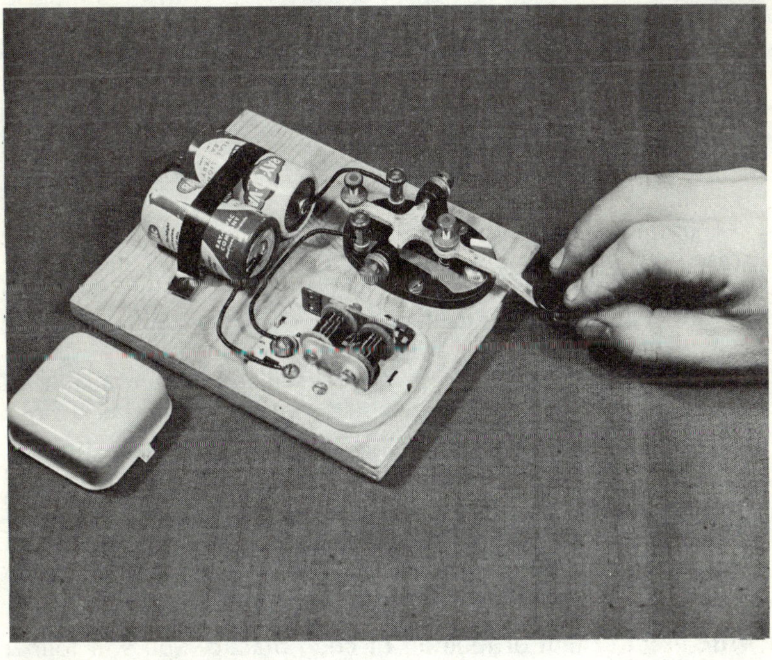

Home-constructed buzzer set. The use of this unit should be followed by intensive listening to "live" stations on a short-wave receiver.

other, and fasten them down with tape or a strap cut from a tin can.

Wire the components as follows: right-hand binding post of key to center post of the right-hand battery; bottom of right-hand battery to center post of left-hand battery; bottom of left-hand battery to left-hand post of buzzer; right-hand post of buzzer to left-hand post of key. Use solid or flexible wire. Clean all ends well, and solder the connections to the batteries.

A radiotelegraph key is merely a single-pole switch. It is called a "key" because it opens and closes the circuit in which it is connected. To operate it, merely place your first two fingers on the knob, with the thumb alongside, and press down to close the contacts. Relax your fingers slightly, and the lever will move up to open the contacts. The spacing between the contacts and the tension of the spring that keeps them open are both adjustable by means of knurled screws. Experiment with the side-bearing screws and be sure the lever moves up and down without binding.

Spring tension and contact spacing are both matters of personal preference. Some people like very light tension and close contacts, but for a beginner it is somewhat better to use a fairly stiff lever and spacing of about $\frac{1}{16}$ inch or slightly less. With these adjustments, cleaner dits and dahs can be made, and the tendency to slur them together will be lessened. As you develop dexterity with the key you will gradually ease the adjustments.

In the early days of radio, keys were made of heavy brass, and an operator had to pound them vigorously. To this day, hams call themselves "brass pounders," and they refer to radiotelegraphy operating as "pounding brass."

An untreated household buzzer makes a raucous racket. To increase its pitch and make it sound more like actual radio signals, insert a tiny matchstick wedge between the vibrating armature and the contact spring attached to it. A drop of model airplane cement or similar adhesive will keep this wedge in place.

Battery-operated, solid-state code oscillators are available at low prices in both factory-assembled and kit form. The kit jobs can be put together in an hour or less. Units of this type are compact, light in weight, readily portable, and have a pleasing tone that closely resembles that of real radio signals. A typical one is shown at the end of this chapter.

PRACTICING WITH A PARTNER

The exact system or method of code practice you will follow depends on whether you will work alone or with a partner. It's a much easier undertaking if two people start together. They can check each other's mistakes and in general accelerate each other's progress. Father-and-son teams are very effective. Many a dad starts with his son just to help him out and show that he's a good fellow, and he ends up becoming an avid ham himself. What usually happens, if you announce to your friends and family that you're learning the code and intend to go on the air with your own radio transmitter, is that too many assistants volunteer their services!

Let's assume that you have one partner. In addition to the code practice set you will need a supply of ruled paper and a soft pencil or a free-flowing pen. Have your partner sit directly next to you so that you can share the code chart and so that you can each watch what the other writes down when the practice sessions get under way.

Notice something odd about this picture of a group of high-school students practicing the code? The girls outnumber the boys four to two! There are many female amateur radio operators of all ages.

Perhaps you have noted that so far nothing has been said about memorizing the code characters themselves. The time to start is now, with the practice set in front of you. Move it in from the edge of the table so that your arm up to the elbow rests comfortably on the table. If the set tends to move around, weight it down with a book. You cannot make clean, sharp Morse characters if the key is not well secured.

Instead of starting with the letter *A* and working on through, learn the characters according to the four groups. *Do not* say to yourself, "*E,* one dot." Merely look at Chart 2-1, note that in

Group 1 the letter *E* is represented by one dit, and tap the key smartly once. The buzzer will emit a short buzz. This is the sound you must learn to associate with the letter. From now on don't even utter the words dits, dahs, dots, or dashes. Keep all conversation to a minimum and avoid distractions.

When you make the first dit, your partner should glance at the code chart, note silently that a single dit is the letter *E,* and write *E* on his paper. Since this is the simplest and shortest of all the code characters, he has to be pretty dense to miss it. If he doesn't have it down on paper within about three seconds, repeat it. The act of transcribing code signals is called *copying,* and the written material or message is called *copy*. The understanding in all practice sessions is that the receiving operator will copy everything he hears.

Proceed down Group 1 as far as the number 5, sending each character two or three times. Keep looking at your partner's paper and check his copy. After sending *E, I, S, H,* and 5 in straight succession twice, mix 'em up a little. Allow yourself about five minutes of this random sending. Now move the practice set over to your partner and let him repeat the performance for your benefit. In all probability you'll copy with less hesitation than he did because you've already heard the signals from your own sending.

Take the practice set back and send *T, M,* and *O* in succession a few times. Again switch with your partner. If you're both of normal intelligence (or better!) fifteen minutes should be enough to impress the eight simple characters on your brain.

Don't try to measure the relative lengths of the dits and dahs. Just make the dahs appreciably longer than the dits so that your partner can't possibly confuse the two sounds. At the beginning you'll probably make the dahs overly long, but with practice you'll find yourself speeding them up.

Practice Words and Sentences

A learner is greatly encouraged when he is able to make words out of the dits and dahs. Therefore, immediately after both of you have mastered the Group 1 signals, start sending the following practice sentences:

Group 1 Practice Sentences

He is Tom
She is his sis
Tessie is his mom
He shoots moths
Its Moses
Hi Tootsie

Meet me sometime
The time is 5
He is a hot shot
She is the most

These sentences are short and necessarily limited in construction because the words consist only of the eight basic dit and dah characters. Allot yourself the first five to send. Tap out the four dits of *H* of the first word of the first sentence, and watch your partner's paper. The instant he gets the *H* written down, follow with *E*. When he has that, wait a fairly long time, say five seconds, to give him the idea that a new word is coming along, and proceed with *is* and *Tom*.

Numerous amateurs have learned the code by listening to the practice transmissions, at a variety of speeds, from W1AW, the headquarters station of the American Radio Relay League, in Newington, Connecticut. The operating schedules are published monthly in the league's magazine, QST, which every ham should read regularly.

"Gee, I got it!" will undoubtedly be his pleased reaction.

No punctuation marks are included with these practice sentences. In informal ham operating punctuation is rarely used, as it usually is not necessary to the sense of the messages. In formal messages, where punctuation is required, the marks are usually spelled out as complete words. The use of the punctuation marks and special characters is discussed later in this book.

Correcting Errors

If your partner copies a character incorrectly, merely touch his arm, shake your head, and repeat it until he does get it right.

What do you do if you send a letter incorrectly? For example, you inadvertently add an extra dit to the letter *S* of the word *is*. This makes the letter *H,* and if your partner copies it as *H* he is perfectly right. Look in the list of punctuation and find the *Errors* signal. This is a string of eight dits. Send it. The technique now is for the receiving operator to cross out the word *ih,* which is what he copied, and to wait for you to repeat the entire word correctly. During all of this, there should be no conversation. The less talking, the faster the progress.

After your partner has made perfect copy of the first five sentences, switch places and let him send the second group of five to you. Allow yourself a total of about an hour for Group 1, and don't attempt the other groups the first night. It's a mistake to cram too much into one session. If you stick to a daily schedule you should have the entire code down pat in a week, and from then on speed is entirely a matter of practice. Remember, copy only what you hear, and don't try to fill out words in advance. You can be fooled very easily!

The second evening, begin by reviewing all of Group 1. With this as a refresher, tackle Group 2, which consists of three characters starting with single dits and three starting with single dahs. Follow the same routine as before, each person sending the letters, first in succession, then scrambled. Now tackle the Group 2 sentences, which contain the characters of Groups 1 and 2.

Group 2 Practice Sentences

The band is not hot
Joe is in Boston
I want to swim
Jane is a new deb
His dish is meat

With the foregoing words you have to pay particular attention to spacing of the signals. If you think you have slurred dits and dahs together when they should be separated, fall back on the error signal and try again. If your partner copies a letter wrong, keep repeating it patiently until he gets it right. If he continues to have trouble with certain letters, break the block by exchanging places and letting him pound brass. It's not unusual for some people to learn the code more quickly from their own sending than from another person's.

Check your handwriting as well as your text during all practice sessions. Here's a good tip: Write small letters. They take less time than big ones, and time is a very important part of the code test.

The third evening, review Groups 1 and 2, and, then progress to Group 3. The practice sentences are now longer and include characters from all three groups:

Group 3 Practice Sentences

Love that roast ham
Lets have some fun
This fish is too hot to eat so soon
His vest is red and blue
The wash is out 5 minutes

It's still too soon to concern yourself about copying speed. There's time for that after you master Group 4 and the rest of the numbers, punctuation marks, and special characters. Then you can use any text from a newspaper or book, and check your speed against a clock. To make a five-word-per-minute check, count off 75 letters in a sentence or series of sentences and send them in three minutes. You'll undoubtedly have to try several different rates of sending before you hit the approximate five-word-per-minute rate. You'll be pleasantly surprised, after a week or so of diligent practice, to learn that 5 wpm is very, very slow.

As you practice, you'll discover that you can understand or "read" whole words and even short sentences without writing down the individual letters as they are transmitted. This is known as "copying in the head." Experienced operators carry on long exchanges of conversation in this manner without putting a word on paper. Practice is what does it.

LEARNING ALONE

If you have more than one eager-beaver partner, follow the foregoing routine exactly as described, but allow a little more time for each session. But suppose you have no partners at all . . . in other words, you are forced to work alone. What to do? Take heart from the fact that thousands of hams have learned the code, read up on the technical questions, and passed the license test entirely on their own and without assistance of any kind. You can do as well if you merely make up your mind to it.

Start with the same buzzer practice set, memorize the code by groups, and maintain a steady pace of daily sending exercises of printed text matter. The more you send the more thoroughly your brain will associate the combinations of sounds with letters and numbers. You'll finally reach a point, of course, where you simply must have a real receiving practice. That is the time to buy a good short-wave receiver—you'll need one anyway for your "shack." (See Chapter 4.)

You can't possibly imagine how many CW stations are on the air, morning, noon, and night, until you start tuning across the various short-wave bands. Not just ham stations, but hundreds of commercial ship and shore stations, naval and military stations, etc. The transmitting at most large stations is not done with hand keys, but with motor-driven tape machines. These grind out perfectly formed and spaced characters that are a pleasure to copy. Tune around, pick out some loud station that sounds like fair game, and try your luck.

For "live" receiving practice on a more organized basis, look up the current operating schedules of ham stations that send accurately timed practice text, at announced speeds, for the specific benefit of beginners. You'll find these schedules in the monthly ham magazines, in general-interest publications such as *Popular Electronics* and *Radio-Electronics*, and even in the club news columns of some local newspapers.

Copy what you can, and don't try to figure out lost characters. You'll hear all sorts of things: clear text press dispatches, personal messages, ship and weather reports, etc., in a variety of languages that use the basic "English" alphabet. There's nothing to prevent you from copying these transmissions, but don't let them go any

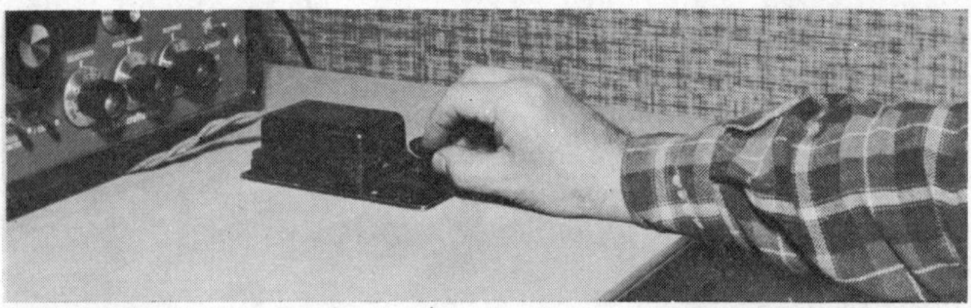

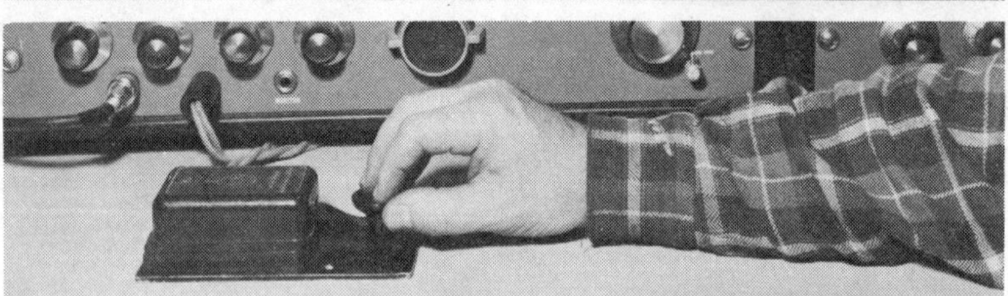

The wrong and right way to use a radio key: In the top photo the arm is too far off the table and will tire easily. In the bottom photo the arm is supported from the elbow to the wrist. This position leaves the fingers and the arm muscles relaxed and is conducive to good sending.

further than your shack. It's unlawful to divulge the contents of specific messages to people other than the addressees. The matter is covered in Section 605 of the Communications Act of 1934, "Unauthorized Publication of Communications." Actually, you'll throw away your practice copy as fast as you accumulate it, but you should know what the law is. The restriction does not apply, by the way, to messages transmitted by hams.

You'll also hear a great deal of "hash" that consists of unpronounceable five-letter groups. This stuff is often referred to as real radio "code"; that is, messages with secret meanings. For hours on end you can copy qwert yuiop lkjhg asdfg zxcvb mnbvc, until you don't know whether you're coming or going. The term "code" in this application is not technically correct. Because there is so much of this type of transmission on the air, the subject can stand a bit of clarification.

CODES AND CIPHERS

Messages having secret or hidden meaning are called *cryptograms*. There are two distinct varieties. In the *code* type (the word "code" having nothing whatsoever to do with the dits and dahs of International Morse), altogether different meanings are assigned to intelligible complete words, phrases, and sentences. This must be done by prearrangement between the sender of the message and the recipient. A message might read, "Oceans of love

Code practice with special records is good for student working alone.

and best wishes for a pleasant crossing," and by itself it makes sense and appears to be innocent. However, the person receiving it might understand it to mean, "Dump the diamonds overboard. Someone has squealed to Customs." For less sinister commercial purposes there exist fat books of code lists whose purpose is economy rather than deception. For instance, the single word "boy" might mean "Arriving in Baltimore," or "girl" might mean "Your order received." When you pay 25 or 50 cents a word for overseas transmission, you want to use as few as possible!

Contrary to the fond belief of many writers of romantic spy stories, there is no way of "breaking a code." You either know the arbitrary meanings of the words or phrases or you don't. What these writers have in mind when they say "code" is properly known as *cipher*. In a ciphered message the individual letters of the original clear text are either rearranged or are replaced by other letters, the result in either case being unintelligible. The method of scrambling must of course be known to both parties, and is called the *key*. A very simple but obvious cipher has the whole message sent backwards. In another, a reversed alphabet is used to replace the letters of the original; that is, *A* becomes *Z*, *B* become *Y*, etc. Through intensive study of the recurrence rates of the various letters, it is possible to "break" even very complicated ciphers.

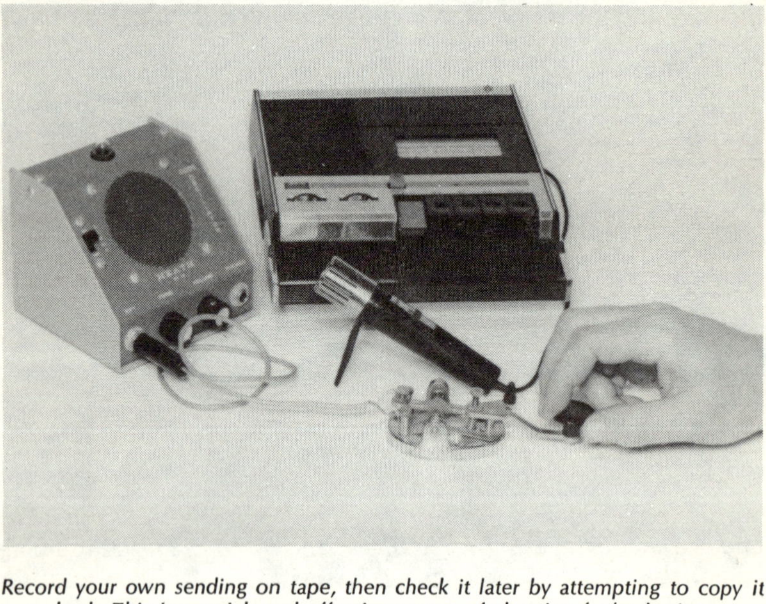

Record your own sending on tape, then check it later by attempting to copy it back. This is a quick and effective means of showing faulty keying.

Cipher messages are usually sent in the form of five-letter groups just to make things harder for cryptographers who want to break them down for one reason or another. If the words are kept in their original lengths it is fairly easy to identify many of them. For instance, in English a single letter word would have to be *A* or *I;* two letters could be *is, am,* be, *in,* etc. When a cipher message in five-letter form is deciphered, the recipient only has to read it slowly to reform the letters into understandable words.

Cipher traffic is rather dull to copy for code practice, but its virtues are its plentitude and its accuracy. A large percentage of it is of military origin and is machine-transmitted.

CODE RECORDS AND TAPES

Phonograph records and magnetic tapes of practice transmissions at various speeds are available at reasonable prices. They are excellent for both individual and group instruction because they can be run repeatedly until the students know them thoroughly; then they can usually be sold—or donated, if you feel magnanimous—to other newcomers.

A tape recorder of any size or type is a very valuable aid to a would-be ham. To record live signals from a CW station, merely place the microphone near the loudspeaker of the receiver. You can then run the tape as often as you wish either for yourself or for friends. You can also use the machine to check your own "fist." Proceed as follows:

Select a column of stock quotations, hog prices, money exchange rates, or similar material containing mixed words and figures that can't be guessed at, from a newspaper. Place your buzzer set or code oscillator close to the recorder microphone and tap off about ten minutes of this text. Rewind the tape and shut off the machine. The next day, play the tape back and compare your copy with the original. You may be in for a rather large surprise. Your first reaction will probably be something like, "Do I really sound like that?"

Another stunt is to exchange tapes with a friend who also has a recorder. To start, ask a lot of questions, and include a few of arithmetical nature. His answers will indicate whether he read you correctly.

Code Instruction System

Over a period of more than 40 years the writer has taught the code to many individuals, small groups gathered around a dining room table, troops of Boy Scouts, radio clubs, and also to much larger groups of GI's. He is convinced that there is no effective "system" of learning other than steady, applied practice. Once you start, maintain your momentum with daily sessions at the buzzer set or your short-wave receiver. If a partner fails to show up one evening, run over some practice sentences yourself. The more you practice, the more quickly you will get into the rhythm of the dits and dahs, and the more firmly the code will become fixed in your consciousness. If at times you feel discouraged because the signals fail to register, remember that children less than ten years old, as well as grandparents in their late sixties, have obtained ham tickets. Their success poses a challenge; surely, you can make it too.

Compact, battery-operated transistor code oscillator made from a Heathkit. Loudspeaker is inside case and phone jack is on front panel.

COPYING ON A TYPEWRITER

It occurs to many would-be hams that it is much easier to copy the radio code on a typewriter than with pencil and paper. It is. The catch is that you must use pencil and paper when you take the FCC code test, so it's better to learn that way. Of course, there's nothing to prevent you from using a "mill," as operators call a typewriter, in your own shack.

3

A Kit Provides Valuable Experience

If you already have a fair technical background and through experience with Citizens Band, hi-fi sound or television equipment, you will not require much preparation for ham radio. All you need, after your license arrives in the mail, is money in the bank!

However, if you are a beginner (remember, everyone was at one time), you should know right away that ham gear in general is sophisticated in circuitry and complicated in construction. In fact, this is part of its appeal to gadget-minded Americans. You must have a pretty fair knowledge of both the theoretical and practical aspects of electronics in order to use expensive receivers, transmitters, transceivers, and related items safely and to maximum advantage.

You can easily read yourself into basic theory with the aid of numerous study guides on the subject, but you can acquire practical know-how only by working on actual equipment. If you don't know the difference between a megaphone and a megahertz, do as thousands of other hams have done: begin with a do-it-yourself electronics kit of some sort. It should be fairly simple, but of continuing usefulness after you finish it. Don't undertake a receiver or a transmitter as a first project; even the relatively simple ones can get to be rather puzzling about halfway through the instruction manual.

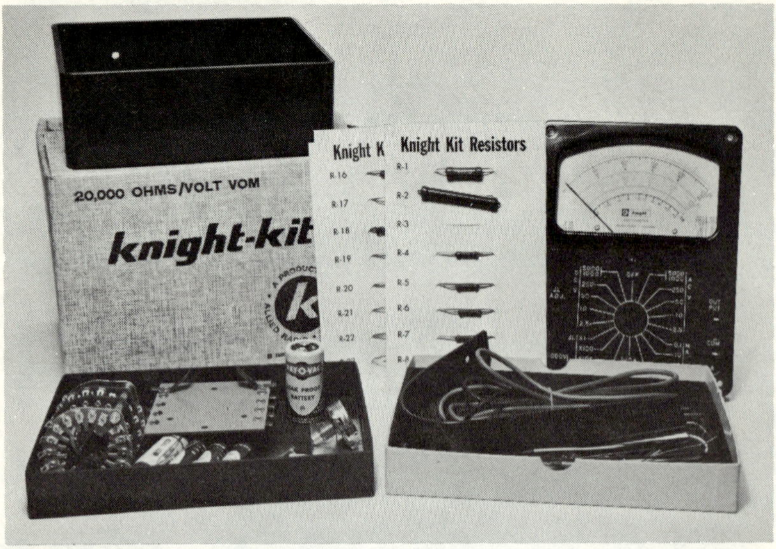

A typical vom kit with its components spread out. Resistors are numbered individually for easy identification. The meter proper is part of the front panel. There are several similar kits on the market.

The chassis, panel, and cabinet of all kits are preformed and predrilled, so assembly of the components is mostly nuts-and-bolts work. You probably already have the few necessary tools, shown later in this chapter, with the possible exception of the

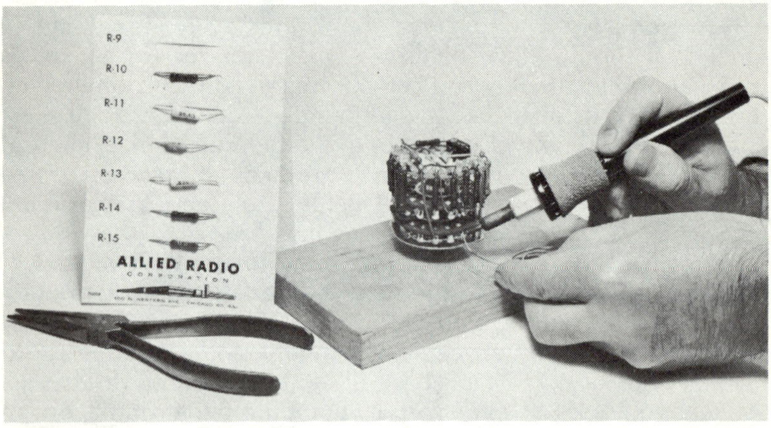

The heart of a vom is a multiposition rotary switch, to which a cluster of small resistors is soldered. Here, the shaft of the switch is placed in a hole in the scrap of wood for support. Note the use of a pencil-tip iron for soldering the leads of the resistors to the lugs of the switch.

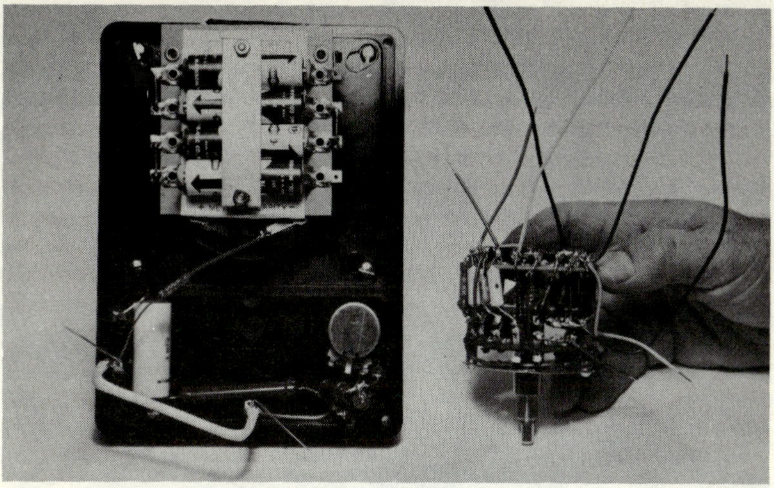

The completed switch assembly for the vom is shown at the right. It is ready for mounting in the lower center hole of the panel at the left. The long wires from the switch will be connected to components on the panel.

wire stripper. A great time and temper saver, this tool removes the insulation from several sizes of both stranded and solid wire by the squeezing action of the arms. The usual method of shaving off

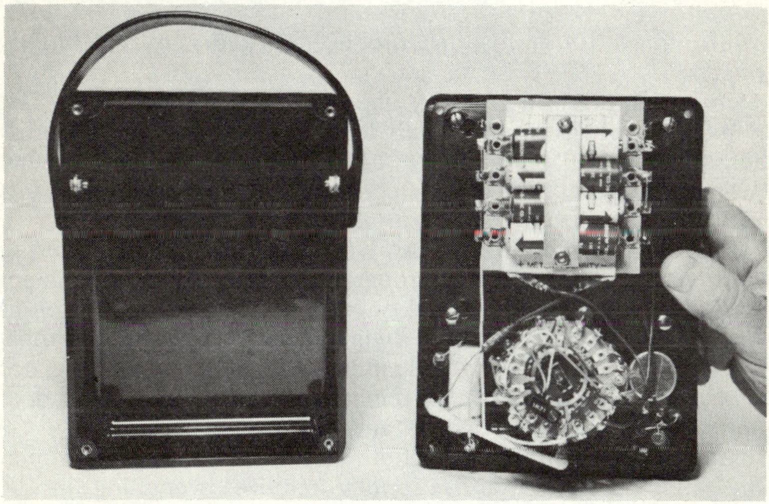

The vom panel at the right is completely assembled. The four penlight batteries at the top are held under a clamp. They last a long time, and are easily replaced. The panel fits sungly in the molded case.

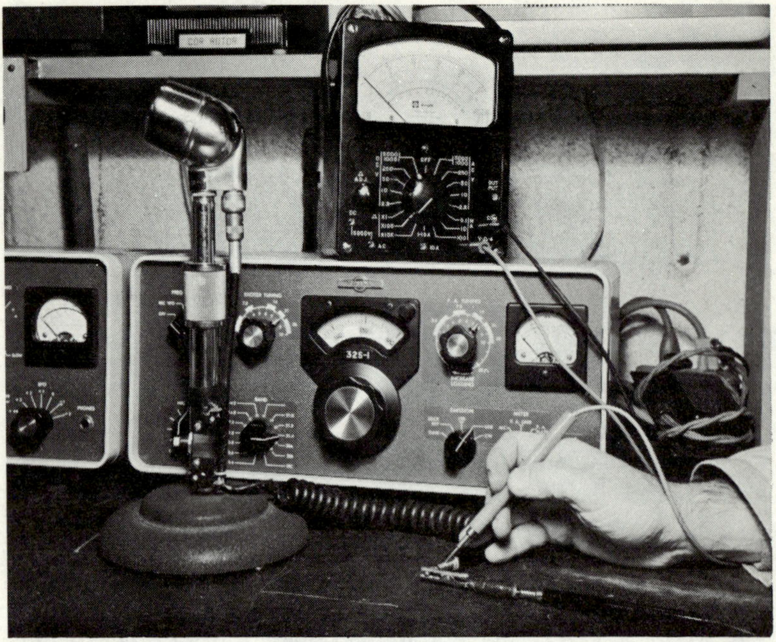

"Continuity checking" is an important use of the vom. Here, the meter probes are connected to a mike plug to test the push-to-talk switch on the stand. If the switch is open, the meter should read infinity ohms; closed, it should read zero ohms.

insulation with a knife often produces nicked wire and bloody fingers.

The soldering iron is a very important tool for kit wiring. Get one of the pencil-tip irons, with replaceable screw-in tips ⅛ inch and 3/16 inch in diameter rated at about 35 or 40 watts. Use only rosin-core solder, which has no corrosive aftereffects. Most kits include a roll of thin rosin-core solder.

For a starter project that can be tackled between sessions of code practice, the author strongly recommends an instrument called a volt-ohm-milliameter, or *vom* for short. See Chapter 10 for a detailed description of the vom. This instrument is absolutely invaluable for a wide variety of circuit and component testing, not only in radio apparatus but also in practically all home appliances and in automobile, motorcycle, and boat electrical systems.

Don't try to complete the entire kit in one evening. First open the shipping carton and carefully remove everything inside. Unfold all packing material and make certain that no small parts are left hiding in it before you discard the packing material. Open all containers, spread the contents on a smooth surface, and check

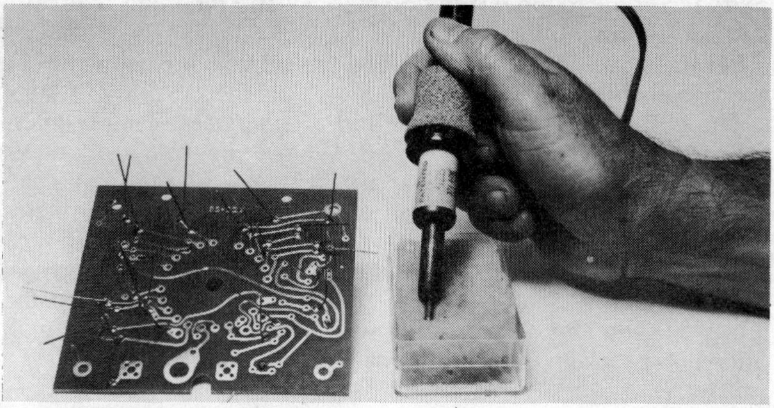

Keep soldering iron tip clean by wiping frequently across damp sponge. At left is a small printed-circuit board of a typical kit project. Note protruding pigtails of parts on other side.

all components against the parts list in the instruction booklet. The instruction booklet often contains drawings or photos of the components to enable you to identify them quickly. A stamp magnifier might be useful for reading the markings on capacitors, resistors, solid-state devices, and other components. Be satisfied with all this as the first phase of the project.

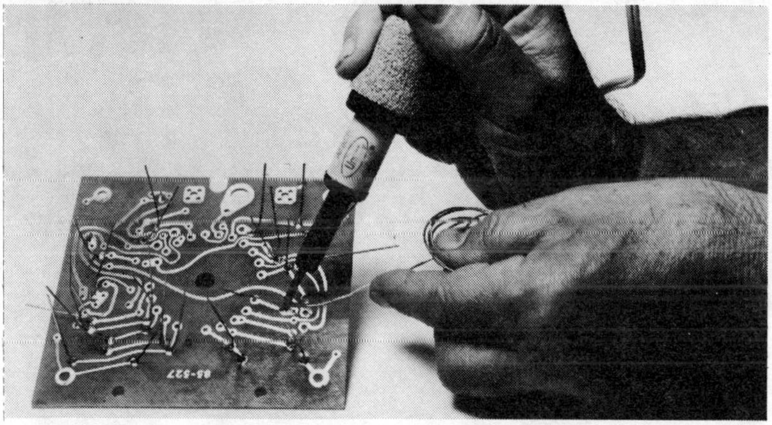

Angle the iron carefully to prevent solder from bridging circuit lines.

The next evening, get acquainted with the soldering operation. Try joining scraps of wire. Observe that the molten solder runs like water. Note that soldering is foolproof with a clean iron, and utterly impossible with a dirty one. Assemble and connect a few parts according to the instruction manual, and then quit. The third day, tackle the job in earnest.

Several stages in the assembly of a typical vom are shown in the accompanying illustrations.

The cost saving between a kit and a comparable factory-made instrument is not really very great. What is important about kit work, aside from its educational aspect, is the relaxation and satisfaction it gives. You forget all about your everyday problems when you concentrate on soldering three resistors and two capacitors into one hole of a terminal lug or guiding a transmitter signal from a crystal oscillator through to the antenna. Also, when the job is finished and working, you will unconsciously puff up with pride and proclaim to all within earshot, "I made it myself!"

KIT TROUBLES

Producers of kits say that there are no *kit* troubles, only *builder* troubles. The writer has visited the service departments of several large kit suppliers, and has examined many finished but inoperative items sent in for adjustment; he agrees with the general diag-

Underchassis view of a typical ham transmitter in process of assembly shows how resistors and capacitors are mounted by their own leads to lugs and terminals. The major part of the entire project is soldering.

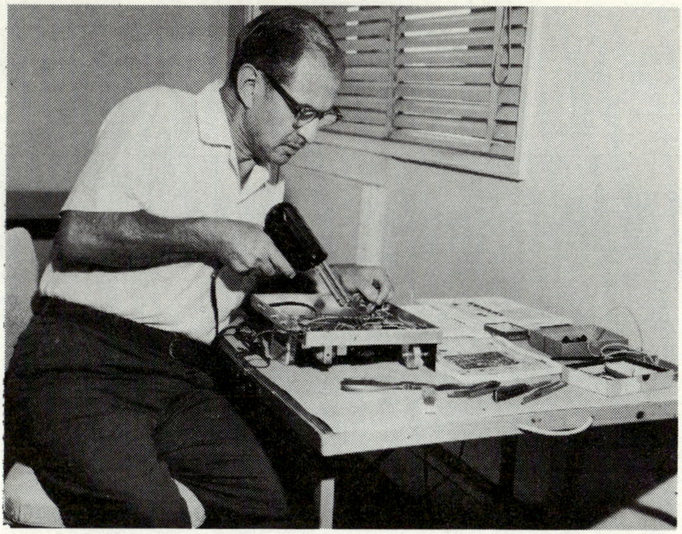

An ordinary card table makes an excellant temporary workbench for kit work. Keep small parts and hardware in shallow boxes to prevent them from rolling off. Hand tool shown here is a soldering gun.

nosis of the repair technicians that poor soldering is the major headache. Builders simply do not keep the irons on joints long enough to cook out the gummy rosin flux. When the flux cools, it usually forms an excellent film of insulation between the surfaces the solder is supposed to unite. Time and again, the mere application of a clean, hot iron to some or all of the connections in a set is enough to bring it to life.

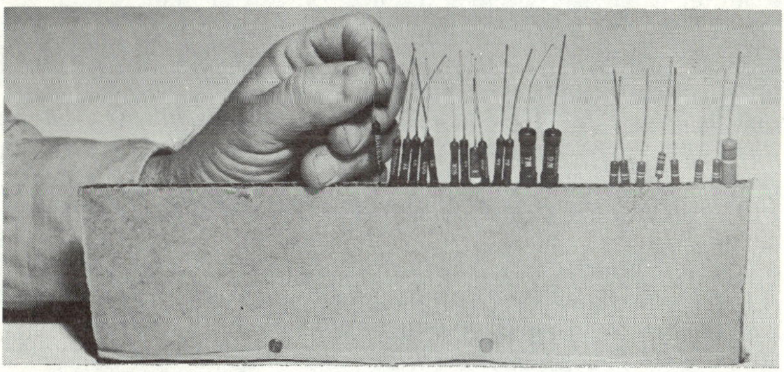

Corrugated board is handy support for small resistors and capacitors. First identify their values from their color coding, arrange them in ascending order from left to right, and stick their pigtails in the board.

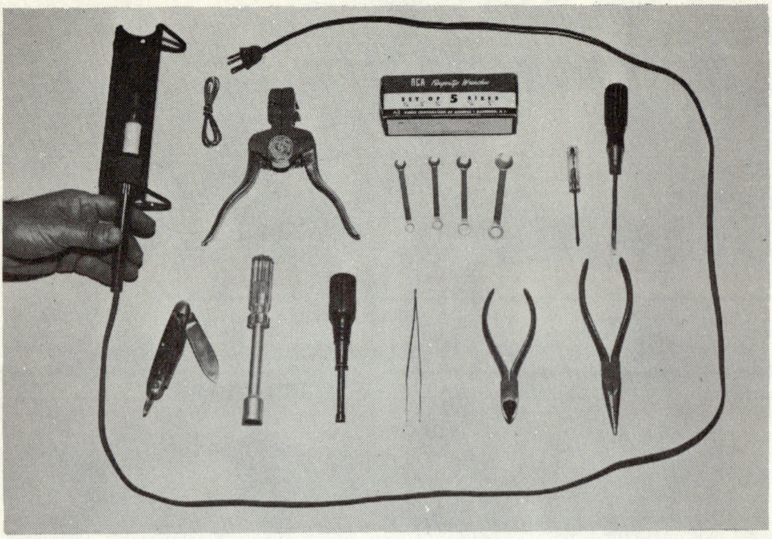

With these basic hand tools you can assemble and wire any kit. From left to right they are: (top row) soldering iron, solder, wire stripper, ignition-type wrenches, screwdrivers; (bottom row) knife, nut drivers, tweezers, side cutters, and long nose pliers. The tweezers are especially useful for picking up and holding soldering lugs, washers, nuts, transistors, etc.

Another problem is the failure of builders to follow the step-by-step assembly and wiring procedures prepared by the kit makers at great expense. The worst offenders in this respect are people with previous experience in electronic construction. They seem to think that these instructions are only for beginners, and that they can bypass them and follow only the schematic diagram and the photos. Some can, but evidently most cannot. Beginners who rely on the instructions and obey them explicitly have a much higher order of success.

A peculiar difficulty that defied isolation for quite some time is color blindness in varying degrees on the part of kit purchasers. Pick up any bunch of small resistors, look at all the color bands, and see if you can name them. To avoid trouble later, it is worth taking the time to check resistors individually on the *ohms* range of a vom.

Polarity markings on electrolytic capacitors are a related problem. If leads are not marked specifically + and − or POS and NEG, look for colored wires or color stripings of some sort. Red is the universally used positive marking and black is negative; any other colors are also positive. In metal-cased units, the can is always negative, all other connections are positive. Some metal- and

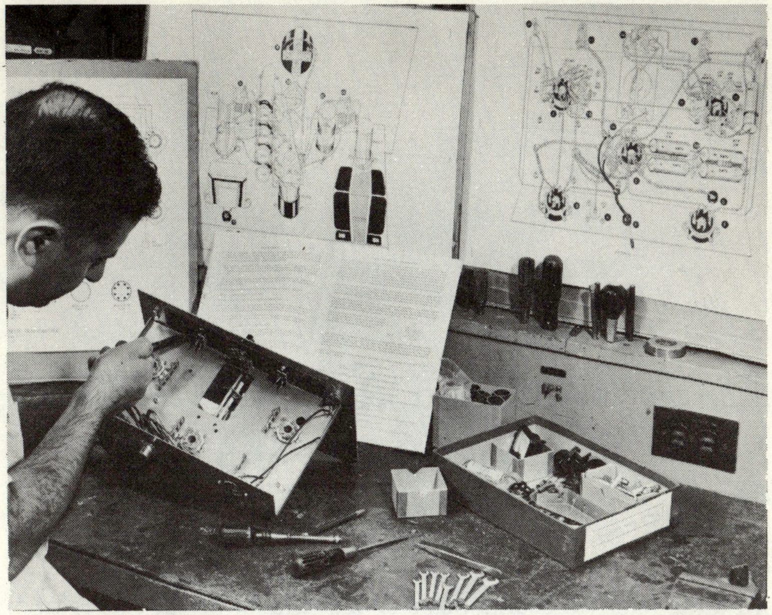

Clip or staple diagrams to cardboard and stand them upright on workbench for easy viewing. Strong but diffused illumination over work area is important. Double 40-watt fluorescent fixture is highly recommended.

paper-cased capacitors have no polarity markings of any type; these units are paper-dielectric units and can be connected into a circuit either way. Likewise, mica capacitors have no polarity markings.

Foolish little things constantly turn up. Failure to put a line fuse in the holder on the back of a chassis is common. Tubes are placed in the wrong sockets. Meters are connected backward. Wires with insulation still on them are twisted together and caked with undigested solder.

The most puzzling cases of all concern equipment that is sent in for "repair" yet works perfectly when hooked up in the shop. Possibly the shaking a chassis receives during the trip breaks off the caking in one or more cold-soldered joints. In at least one instance it was learned—only after a considerable exchange of correspondence with the customer—that the wall outlet into which he had plugged his receiver was dead. Beat that!

4

A Good Receiver Is Worth Having

The first time you look at communications-type receivers you are quite likely to experience a shock—the kind that comes from handling price tags, not bare high-voltage wires. For about as much money as you'd pay for a small color tv set, a window air-conditioner, a refrigerator, or a drill press, you get a rather severe looking piece of equipment that doesn't even crowd the top of a card table. What's more, the speaker usually isn't included, but costs extra!

It isn't until you look inside and consider the capabilities of these sets that you begin to understand why they cost as much as they do, why they are excellent investments, and also why they sometimes outlive their owners. Communications receivers are designed to meet performance requirements entirely different from those of conventional broadcast and television receivers. From the technical standpoint, they bear as little relationship to the latter as a Jaguar does to a Model A Ford.

The frequency allocations for tv and sound broadcasting stations are set up so that in any one center of population there is virtually no mutual interference. The transmitters themselves are powerful (50 kilowatt sound stations, for example, are common), so perfectly satisfactory reception is possible with small, inexpensive sets of low sensitivity and selectivity. On the other hand, the ham bands are wide open for unlimited use by thousands of licensees, the great majority of whom use only a small fraction of the one-kilowatt maximum power permitted by the FCC. To separate the weak signals of a dozen different ham transmitters—

all seemingly coming in on the same spot on the dial—a receiver needs lots of circuits, lots of tubes or solid-state devices, and lots of refinements. Sure, these cost money, but once you've paid for them you have something to enjoy for a long time. It is virtually impossible to "wear out" a good receiver.

The Drake DSR-2 is a sophisticated general-coverage short-wave receiver, covering 10 kHz through 30 MHz. It has a digital readout of frequency accurate to 100 hertz, displayed along top edge of front panel.

RECEIVER CHARACTERISTICS

A number of things make a "good" receiver. It isn't easy to express these qualities in quantitative or even relative terms, because methods of measurements and the interpretation of the measurements themselves do not conform to really rigid standards.

In the consideration of receiver ratings or performance promises, an important factor often overlooked is the skill of the operator. It takes weeks to master the maze of controls on some of the more advanced receivers. Even some of the simple jobs have as many as a dozen knobs and dials, all of which do something to the reception.

So that you might be able to understand something about receivers and evaluate them in your own mind, let's take a broad general look into the various elements of receiver design and construction.

Sensitivity

Sensitivity is the measure of the ability of a set to pick up weak signals. Sensitivity, amplification, and gain all mean the same thing and are used interchangeably. It is relatively easy to build up sensitivity by using enough tubes or solid-state devices as amplifiers. The first stage gives the signal a little boost, the second strengthens

The Yaesu FR-101 covers not only the ham bands from 2 through 160 meters, but also practically all of the short-wave broadcasting bands. It uses plug-in circuit boards that greatly simplify servicing and inspection. Completely solid-state, this set contains 20 transistors, 12 field-effect transistors, 4 integrated circuits, and 33 diodes.

it some more, and so on along the line. Present-day circuits and tubes are the highly refined results of more than 50 years of intensive research and development. There is some feeling in engineering circles that the ultimate in usable sensitivity has been obtained from most vacuum tubes and related transistors. In fact,

The Drake SSR-1 is a modest all-band communications receiver with built-in speaker, direct-reading tuning dial, signal-strength meter, ssb and am capability, and coverage to 30 MHz.

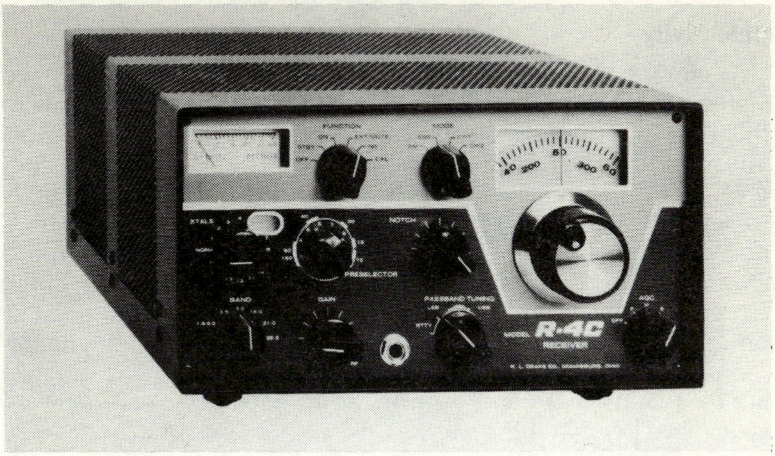

The Drake R-4C tunes the 15-, 20-, 40-, and 80-meter ham bands and the 28.5–29.0-MHz section of the 10-meter band. It also covers 160 meters with an auxiliary crystal. This receiver can be used for Military Affiliated Radio System stations, CB, WWV time signals, foreign broadcasts, etc. Power and loudspeaker are accessories.

sensitivity claims by themselves do not mean much anymore unless they are qualified in terms of signal-to-noise ratio. Some amplifier arrangements are so extraordinarily sensitive that they reproduce, as audible sounds, the slightest irregularity of electron emission from the cathodes or filaments of the tubes. This sound is usually heard as a high-pitched hiss. If a weak radio signal entering these tubes has electron movement of the same very low order, the signal and the noise are both amplified and reproduced equally, and the signal usually loses its identity as such in the output.

Often a ham condemns a set as "noisy" when the interference is actually external and not internal. In some industrial areas, particularly those fed by overhead high-voltage power lines, the air is literally saturated with strong electrical impulses. In dry weather, at night, a blue discharge around the line insulators is clearly visible (the effect, eerie indeed, is called "corona"), and the disturbance in the receiver takes the form of loud crackling. In wet weather the noise is more of a frying sound.

Solid-state devices were slow to replace time-tested tubes in receivers because early transistor types were inherently noisy, offered low gain, and were difficult to manufacture with uniformity. However, all of these problems have been solved. Practically all receivers of recent design are entirely solid-state and have established high reputations for their overall performance.

Selectivity

There are over three hundred thousand hams in the United States alone, and there are times when you'll swear that half of them are transmitting on the same frequency at the same time. The most important feature of a modern amateur communications receiver thus is its selectivity. This is defined simply as the ability to separate signals of closely adjacent frequencies. A set with a high order of selectivity is said to be "sharp" or to "tune sharply." One with poorer selectivity is "broad," or "tunes broadly." In the better ham receivers the selectivity is adjustable from relative broadness to needle-point sharpness, to suit the skill of the operator and the interference conditions of the moment. The greatest improvements that have been made in ham receivers in recent years have been in this area.

A remarkable degree of sharpness is achieved in some communications receivers through the use of adjustable or fixed filters of the electrical, mechanical, and piezoelectric (quartz crystal) types. For many years the crystal filter was considered the most effective. However, the trend is now away from crystals to the more flexible electrical filters.

Bandspreading

A means of spreading out the ham bands so that they occupy a large portion of a tuning scale instead of only a few degrees on it is a basic feature of all amateur receivers. Bandspread plus good selectivity enable you to crawl slowly across a band and to unscramble what at first sounds like a mess of stations.

In most sets bandspreading is accomplished by the use of a small multiple-section variable capacitor connected in parallel with the main tuning capacitor and equipped with its own calibrated dial. Other electrical means, sometimes combined with dials having high reduction drive ratios, are also employed. To facilitate tuning, which can be very critical, the controls must be smooth, easy, and absolutely free of backlash.

Digital Frequency Readout

Probably the most eye-catching feature of the solid-state receivers and transceivers is digital readout of the frequency on which the equipment is working. (A digit is any number from Ø to 9, and is so named because long ago people used their toes and fingers for counting purposes).

In a typical arrangement, electronically energized digits about ¼-inch high (or larger) appear in a window just above the tuning knob. Suppose we start with an indication of 14.200.0 megahertz.

A mere touch on the knob changes the digit 0 to 1, for a frequency of 14.200.1. The last 1 actually represents 100 hertz, a fantastic accuracy when you consider that this is only one hundred parts out of fourteen million! Jiggle the knob again in the same direction and the numbers flicker in progression: 14.200.2; 14.200.3; 14.200.4; and so on until the last number is 9, for 900 hertz. The next turn of the knob gives a readout of 14.201.0 MHz; that is, the frequency has gone up 1000 hertz, or one kilohertz. Turning the knob in the other direction merely causes the numbers to go down instead of up.

Initially, digital readout was an accessory in the form of a little box that was placed on top of a receiver or transceiver and connected to the chassis by a small cable. In some sets, space for the unit was provided so that it could be dropped in and connected quickly with no external wires showing. Digital display proved so popular, in spite of its additional cost of about $100, that manufacturers now include it as original equipment in many of their models. It doesn't make a set work any better, but it certainly makes it more fun to use!

Stability

If you have to retune a receiver slightly every now and then to hold the signals of a station, either the receiver or the transmitter can be suffering from lack of stability. This drifting action in receivers is usually due to changes in the values of capacitors and other parts as they absorb heat from nearby tubes, transformers, and resistors. The problem has been pretty well licked by the use of various circuit refinements and special temperature-compensated capacitors in certain critical circuit positions.

It is normal for a receiver to drift a little after it has been turned on. Sets using tubes take as long as 30 minutes to settle down, and they might need a touch up even after that to reach complete stability. Some operators keep tube-type sets warm all the time, with the volume control turned down to silence them when the station is off the air. Tube-type receivers often have a STANDBY switch which accomplishes the same thing.

The problem of stability scarcely exists in newer receivers of all solid-state design. In a typical solid-state set, a frequency readout might flicker up and down 100 hertz for only 10 or 15 *seconds* before it locks firmly in place.

If some stations stay put on your dial and others seem to wander, you are safe in assuming that your receiver is stable and that the transmitters are at fault. Don't be confused by fading. This is a change in the strength of a signal, but not in its dial setting. All you can do to catch a fading signal is to crank up the volume

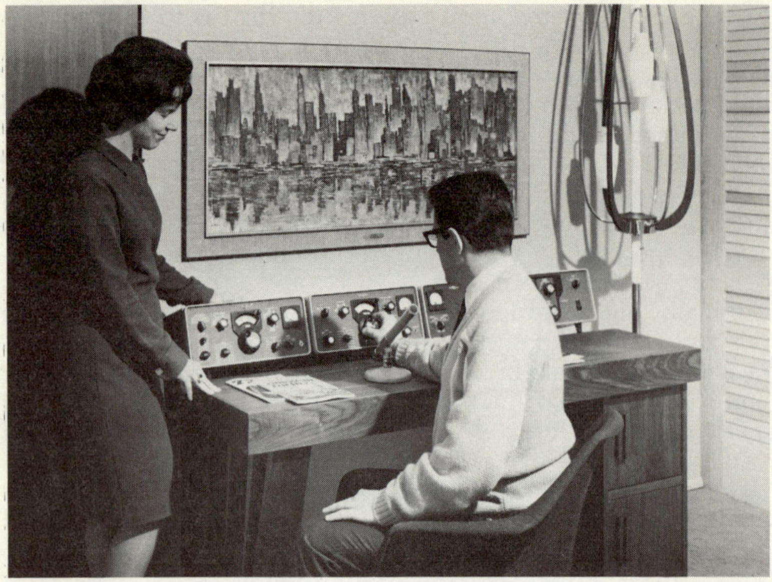

"Wife approved" is how Ray Grenier, K9KHW, describes this handsome operating console in the living room of their home. Desk is the Design Industries "Diplomat" obtained from Amateur Electronic Supply, Milwaukee, Wisconsin. Back edge is angled to put panels of Collins "S" equipment directly in line with operator's vision.

control to maximum and hope that the fade will reverse its direction.

In actual operating practice, drift is much less serious than it sounds. If a signal does wander a little you can usually restore it with a barely perceptible touch on the tuning knob of the receiver.

FREQUENCY COVERAGE

There are two distinct types of communications receivers: the "general-coverage" or "all-wave type," and the "ham-bands-only type." The tuning range of the first is extremely wide. It usually starts at 540 kHz (the bottom of the broadcast band) and runs upward without gaps to 30 MHz, which is beyond the top of the 10-meter amateur band. Thus, it includes the ham bands from 160 through 10 meters in addition to an enormous variety of other short-wave services. In some sets, separate bandspread scales are provided for the ham bands; in others, any portion of any band can be spread out for easier tuning. The one advantage of a general-coverage receiver is its versatility. When it is not being used for hamming, it can furnish all members of the family unlimited listening possibilities on the police, fire, citizens band, aeronautical, maritime, and domestic and international broadcasting channels.

The ham-band receiver defines itself. It is a highly specialized set offering superior selectivity and bandspread on the amateur channels, and on only some of them at that. Most models take in only the 80-, 40-, 20-, 15-, and 10-meter bands, which are by far the most popular bands in general use; others might include the 160-meter band and the 15-MHz frequency of certain time-signal stations.

The Yaesu FRG-7 was designed specically as a stand-by receiver for hams. It is a general-coverage receiver giving high performance of frequencies from 550 kilohertz to 29.9 megahertz. It is all solid-state with a built-in loudspeaker.

Both types of receivers sound very desirable. Which one to buy? Before you buy either, consider a third choice: a single-unit transmitter/receiver called a "transceiver," which is built into a cabinet about the same size used for separate transmitters and receivers that perform the same functions. This is possible because of the tremendous strides made in recent years in the miniaturization of electronic equipment of many kinds. In its receiver mode a transceiver does everything a separate ham-band receiver does, so you can have live code practice and acquaint yourself with operating procedures while you're studying for the exams. Then, when your ticket arrives, you are all prepared to go on the air.

If your preference runs to a ham-band receiver, be sure to pick one for which there is a matching transmitter from the same manufacturer. This is an excellent pairing, and offers the advantage of very flexible choice of operating frequencies.

The ideal station set-up would consist of either a transceiver or a matched transmitter and receiver pair, for primary communication purposes, and a general-coverage receiver for casual scanning of its numerous bands.

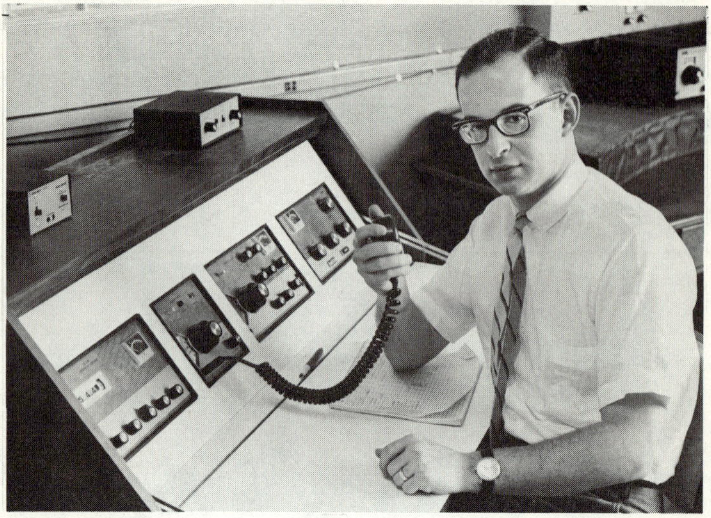

A radio executive whose work is also his hobby is Larry Meyerson, WØWOX, president of World Radio Laboratories. He operates from this business-like console, which contains (left to right) the Galaxy DeLuxe Accessory console, Galaxy *Remote VFO, Galaxy V Mark 2 transceiver, and Galaxy Model 2000 Linear Amplifier. A very neat rig!*

Very few conventional receivers or transceivers include the 6-meter band. The 6-meter is not as widely used as the 10- through 160-meter bands because it is highly vulnerable to changes in atmospheric conditions. However, two types of equipment are available for the 6-meter amateur band. The first type includes relatively simple, low-power, self-contained transceivers for this band alone. The other type is the "transverter" that works in conjunction with a regular transceiver. Operation on 6 meters is something of a sporting proposition, which is why some hams like it.

On all the ham bands other than 2 meters, DX chasing (long-distance communications) is the major activity. On 2 meters, the objective is reliable short-distance (line-of-sight) voice communication, mostly from cars and other vehicles, with flea-power "walkie-talkies" or more powerful (10-watt!) transceivers used for both mobile and base stations. This sounds like CB, but it is far removed from it. See Chapter 12 for more details.

Band Changing

In modern communications receivers, band changing is accomplished by a front-panel switch that connects various tuning coils, capacitors, or crystal oscillators into the circuits. Plug-in coils which must be shifted in and out of a single socket are used in some very simple sets sold in kit form. These receivers are intended primarily to give newcomers a little practical experience in assembly and wiring. While they are quite sensitive and pull in many stations, their selectivity leaves much to be desired, and they cannot be regarded as suitable for anything other than beginner CW operation under favorable conditions.

RECEIVER ACCESSORIES

Oddly enough, the lower priced ham receivers have built-in speakers, but the high priced ones do not! In the latter, the chassis space is taken up with too many more important components.

In communications work, high-fidelity sound reproduction is not only unnecessary but is actually undesirable. We are interested only in the voice range of frequencies, and anything lower or higher is just so much interference. For this reason, the audio amplifier section of ham receivers is usually very modest, and any small speaker (between four and eight inches in diameter) is entirely satisfactory. Most receiver manufacturers sell speakers in small cabinets to match the sets.

Oversize earphones like these have foam-rubber cushions that shut out room noises and permit a radio operator the concentrate on weak signals. Hams commonly refer to these as "cans."

If the ham shack is in the living quarters of a house or an apartment, the other members of the family may be annoyed by the rather noisy mixture of CW and phone signals, static, and ignition interference coming from the speaker. To keep peace, it is advisable to use a pair of earphones, or "cans." In all receivers, the speaker is cut off automatically when the earphone plug is inserted into the earphone jack.

A large variety of earphones is available. The conventional type consists of two units on a double-wire frame, fitting over the head. Light phones of the hearing-aid type are becoming popular too.

Earphones have advantages other than household quietness. Even many experienced technicians do not realize that they are probably the most sensitive of all electronic reproducing devices—they respond to infinitesimal changes of current. Earphones make signals ordinarily lost in a speaker intelligible. Because they require so little energy, you will find you can turn the receiver gain controls way down from the usual adjustments required when a speaker is used. This usually reduces the noise level considerably and makes for more comfortable reception.

A Drake 2-B ham-band receiver is feature in this compact L-shaped radio shack of Mrs. Eileen Cline, KØILM, Fort Madison, Iowa. An expert CW operator, Eileen often flabbergasts other hams by sending at 60 words per minute, a speed that she can follow herself when she is receiving.

Phones put the sound directly into your ears. Fitting snugly against the latter, they shut off room noise and enable you to concentrate better. The more you use phones, the more you will like them and will marvel at their ability to produce weak stations out of nowhere.

RECEIVING ANTENNAS

Short-wave radio signals have amazing penetrating ability, and can be heard quite well with small antennas inside seemingly tight buildings. In fact, sometimes they can be heard with no antenna at all! Evidently they seep in through openings in the metal chassis or cabinet of a receiver, or enter by way of the ac power line. (In the early days of broadcasting, a popular stunt was to connect the antenna lead of a set to one side of the line, with a small mica capacitor in series to eliminate shock hazard.)

The Galaxy R-530 is a rugged solid-state receiver of modular construction. Tuning coverage is continuous from .5 through 30 MHz. It features very accurate frequency readout on the main tuning dial. Really a professional set, which is why many hams like it.

Indoor antennas are a challenge to a ham's ingenuity. You'll probably want to try one immediately after you unpack your new receiver or transceiver. The only supplies you need are a box of pushpins or thumbtacks and about 50 feet of stranded copper wire, preferably size No. 22. As a matter of fact, *any* wire, copper or aluminum, of any size would also work, but the No. 22 is recommended because it is very thin and is easily concealed. The wire can be bare or insulated, whichever is most readily available.

Start with one end of the wire connected to the antenna terminal of your set. For this job you might need either a small phono plug or a screw-on coaxial connector, obtainable from Radio Shack, Olson, or other electronics parts suppliers.

Now, for a starter, spread the wire along the baseboard of the room, fastening it down every five feet or so with a pushpin or thumbtack. When you've rolled out as much wire as you can, tie up the excess and leave it hang. The results you get with such an indoor antenna are likely to surprise you. You're almost sure to hear something, especially during daylight hours on the busy 20-

meter band. In any case, try stringing the wire in other places to see if improvement is possible: hang it along a hallway, diagonally across the ceiling, or on a picture-hanging molding, if available.

Almost any metal body or surface picks up radio waves. Metal window frames and window screens made of aluminum mesh (not nylon) are good. Even bed springs and porch furniture have been used successfully.

Unusually wide frequency coverage is the feature of the Drake SPR-4 solid-state communications receiver. It can actually be programmed for a total of 24 bands, from 150 kHz to 30 MHz. It can be operated from ac or battery; speaker is separate.

If you live in a private home or in an apartment house with an accessible roof you might want to try an outdoor antenna. For this you need a strong antenna wire called "7/22," which consists of seven strands of No. 22 copper wire, and a couple of small glass or ceramic insulators. Hang the antenna as high and as clear from obstructions as possible, with an overall length of between 50 and 100 feet. Take particular care that the antenna won't hit an exposed power line if it comes down during a storm. Eventually you will use a single antenna for both receiving and transmitting, but this temporary antenna will be very satisfactory until then.

QUALITY AND PRICE

It is fortunate for prospective purchasers of communications receivers that the market is remarkably free of "lemons." The excellent reputations enjoyed by ham manufacturers, some of whom have been in the business for half a century, are based on one thing—performance. A manufacturer who wants to stay in

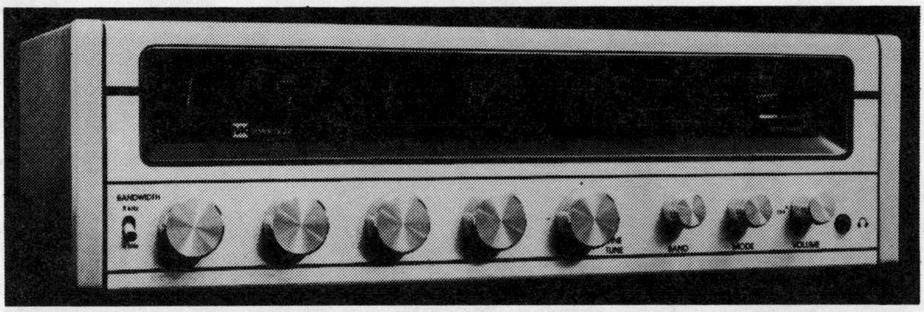

This sleek piece of equipment is the McKay Dymek DR22, a solid-state general-coverage receiver that has digital readout not only on the usual bands between 500 kilohertz and 29.7 megahertz, but also in the interesting and relatively unexplored low-frequency region between 50 kHz and 500 kHz. The circuitry is highly sophisticated and offers a high degree of selectivity with a low noise level.

business cannot afford to put out poor equipment. If he is foolish enough to try, the first unhappy buyer will scream bloody murder over the air, and in a couple of days the whole ham fraternity will know about it. After all, isn't radio the fastest way to disseminate information internationally?

Receivers range in price from $50 to $2200. Why such a great difference in price? The answer is simple. Some people have very little money to spend on their hobbies, others have a moderate amount, and still others hardly look at the price tags. Manufacturers merely try to satisfy them all. Even the lowest priced set enables a ham to enjoy the thrills of local and foreign contacts. The higher priced receivers merely offer more in the way of selec-

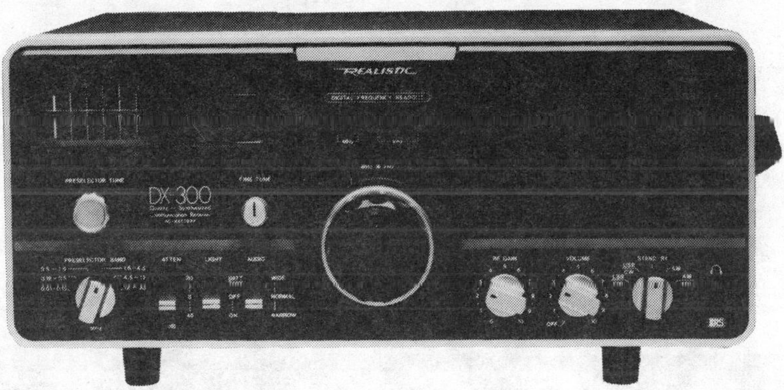

With frequency coverage starting at 10 kHz and going all the way up to 30 MHz, Radio Shack's DX-300 is a general-coverage receiver of unusual capabilities. With it you can hear foreign broadcasts, marine single-sideband (ssb) stations, standard am broadcasts, WWV time signals, ssb and CW ham stations, am/ssb Citizens Band, and aircraft weather reports. It features digital readout of all frequencies, built-in speaker, and also a built-in code-practice oscillator. It operates on batteries or ac.

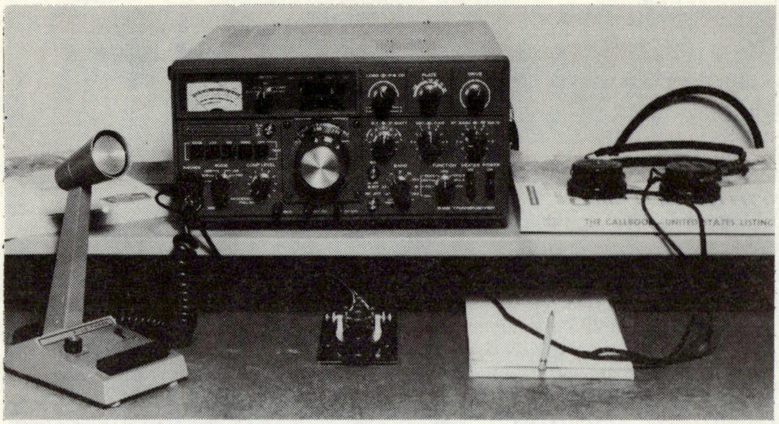

The basic building blocks of this station are a Kenwood TS-820S transceiver, a microphone, a key, and a headset.

tivity, gain, bandspread, accuracy of calibration, ease of tuning, etc.

Communication receivers are primarily built for performance and have little useless ornamentation or other "eyewash." It is therefore possible to grade them directly by cost and to say that two sets of about the same price can be expected to produce

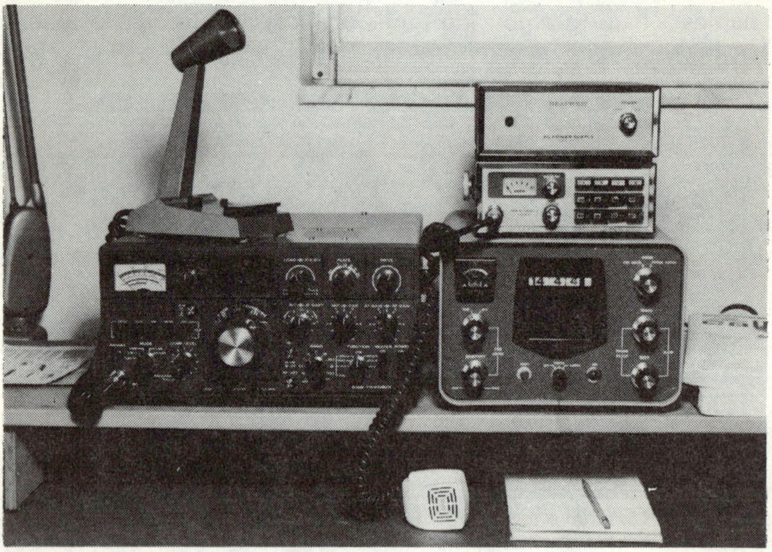

Added later are a combination antenna-meter/phone-patch/clock and a Heath 2-meter transceiver with ac power supply.

about the same results. In general, hams find it wise to invest in the higher priced receivers. Not only are they superior in obvious respects to the cheaper sets, but they have much higher trade-in value. This is an important consideration, because hams by nature are gadgeteers and sooner or later they change their equipment just to have something different—not necessarily new, but just different—to play with.

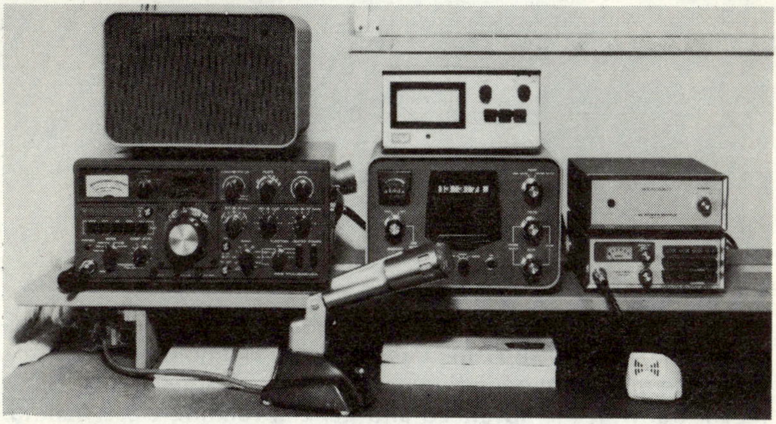

Final additions are a loudspeaker for the transceiver, beam control (above clock), and a new mike. Note convenient shelf arrangement.

POWER CONSUMPTION

Solid-state receivers and the receive sections of most transceivers require so little current that some manufacturers don't even include actual power consumption figures in their spec sheets. A good average with ac operation is about 25 watts. Many sets work directly off 12-volt storage batteries, and some can be run on a power pack consisting of eight flashlight batteries.

A beginner in the ham game is smart if he starts with a few bare essentials and adds to them as his operating skill improves. The preceding photographs show how one man made his station grow.

5

Getting Your Ticket

Of the five classes of amateur licenses, Novice, Technician, General, Advanced and Amateur Extra, you can hold only one at a time. The maximum operating privileges are afforded by the Amateur Extra, with the advanced and general running close behind.

While you are studying the code and radio theory, in preparation for a license examination, by all means write to the Federal Communications Commission, Washington, D.C. 20554, and ask for a copy of the bulletin entitled "Amateur Radio Service." This is extremely valuable because it contains current information on license fees, changes in license requirements and operating privileges, location of FCC facilities, and similar matters of direct interest to would-be hams. The bulletin is free. To make sure that it will reach you, clearly print or type your name, full address, and Zip Code.

As mentioned in Chapter 1, practically anyone is eligible to obtain an amateur radio license, including aliens who have valid United States addresses and are not representatives of foreign governments.

In 1977 and 1978 the Federal Communications Commission promulgated the most important changes in the amateur license regulations since the service was established on an official basis in 1912. Basic requirements are the same as before: code speed of 5 words per minute for Novice and Technician, 13 words per minute for General and Advanced, 20 words per minute for Amateur Extra, and a written test depending on the license class you are trying for. The written exam might include questions on FCC rules, radio phenomena, operating procedures, circuit compo-

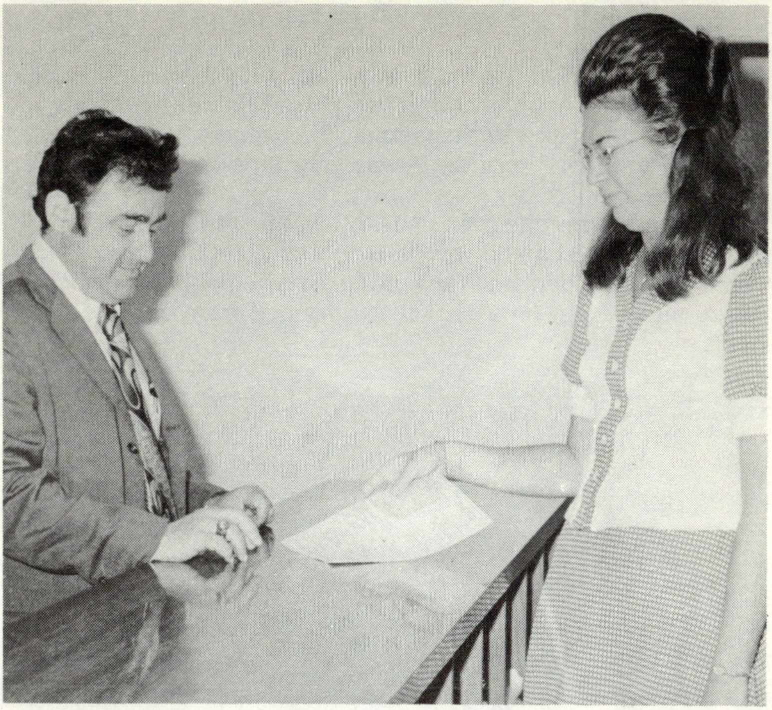

At many FCC offices, applicants for the ham license examination are greeted by affable receptionists who put them at ease with friendly words and smiles. They hand out FCC Form 610, "Application for Individual Amateur Radio Station and/or Operator License," and explain the test procedure.

nents, antennas, etc. The questions are generally of the multiple-choice type.

THE NEW CODE TEST

The code test has been modified greatly. Previously, an applicant had to copy at least one minute of a five-minute transmission without error. Under the new "comprehension" system the tape still runs for five minutes, but you have a choice of three options while you listen to the transmission: (1) you may transcribe every word of it on paper, (2) you may jot down only a few bits and pieces, or (3) you may sit back and "copy in the head" and depend on your memory to get you by the last step. This is a ten-part questionnaire of the multiple-choice type related directly to what you heard from the tape. The passing score is 80%; that is, you need eight correct answers. The FCC examiner looks at your test paper and quickly gives you the verdict. If you pass, you proceed to the written exam.

Formerly, you had to send a few words with a straight key connected to a tone oscillator, after you passed the receiving test and before you took the written exam. This requirement has been dropped, probably because it was very time-consuming with a roomful of applicants.

When the comprehension code test was first tried, some would-be hams protested that it was "too different" or "too hard." However, some extended trial runs produced a surprising statistic: the failure rate of first-time applicants dropped from about 50% to 20%!

This compact solid-state keyer is used at many FCC examination locations, replacing the heavy machines in use for many years. The perforated tape is the same as before, as is the accuracy of transmission. Its built-in speaker eliminates the need for individual earphones, and enables FCC examiners to set up test facilities in any convenient office.

HOW TO PASS THE TESTS

Practice and study are the basis of success, of course. But if you become afflicted with buck fever when you walk through the doors marked "Radio-Operator Examinations," you're only human. You know what buck fever is. All summer long a sportsman will practice rifle shooting. He'll get the trigger squeeze just right, he'll learn how to hold high or low to make the bullet hit where he wants, etc. And what happens when he goes into the woods and sees a big deer in the flesh for the first time? He jerks the trigger, he holds a foot over the target, he misses completely, and then collapses to the ground into a state of nervous exhaustion. That's buck fever!

The same thing happens every day at FCC examining offices. Usually it's nervousness, but sometimes it's lack of preparation. That's the opinion of the radio license examiners in the New York

Relax when you set down for the code test. Under the new "comprehension" plan you only have to get the general sense of the transmission, and then answer questions based on it. However, it's a good idea anyway to transcribe most or all of it, because you are permitted to refer to your copy during the written test.

City office of the FCC, probably the busiest in the country. These examiners process as many as 500 applicants a month, ranging in age from 10 to 72. Most of the examiners have taken the test themselves and are active hams. They are therefore qualified to give some good advice. Here is what one of the examiners told the author:

Instant recognition at club meetings, conventions, picnics, etc., comes from wearing call-letter plates like these. On the bottom one, B.A.R.C. stands for the Broward Amateur Radio Club. Broward is a county in Florida, a relief map of which is to the left.

"Too many people rush down the first time someone checks them off at what they think is 13 words per minute. They don't make enough allowance for timing errors, or the fact that they will take the test in strange surroundings. They should protect themselves by becoming really proficient at full 15 per minute before they try our 13. We don't depend on uncertain hand sending. We use an automatic tape machine that is periodically checked for timing accuracy. When it's adjusted for 13, it sends at

13, no more, no less. When a failing applicant grumbles a little and infers that the sending sounded sort of fast, we just smile and wish him better luck next time."

THE NOVICE LICENSE

The first thing you must learn about the test for the Novice license is that it is not given at FCC offices by FCC personnel. You take it under the supervision of a volunteer examiner, who must be 18 years old or older, the holder of a General, Advanced, or Amateur Extra license, and not related to you. Why isn't the Novice test given by the FCC? Because this agency is swamped with the work of handling literally millions of license documents for a large variety of radio stations and operators. The volunteer program has been very successful and will undoubtedly be continued indefinitely.

How do you find a volunteer examiner? Ham radio clubs are the best sources, and there are many of them. For a list of clubs in your vicinity, write to the headquarters of the hams national organization: The American Radio Relay League, 225 Main Street, Newington, Connecticut 06111. Include a stamped and self-addressed envelope, with your name and address typed or written clearly.

The procedure is as follows: First, you obtain an FCC Form 610, in person or by mail from any FCC office or from a dealer in ham equipment. Look for the latter in the yellow pages of your telephone directory. Take the comprehension 5-word-per-minute code test. For this purpose the examiner will probably use a small cassette player with a prerecorded, accurately timed tape. Within ten days of passing this test, fill out all applicable parts of the Form 610. Send this, with a written request from the examiner for the written exam, to the Federal Communications Commission, Gettysburg, Pennsylvania 17325. In return, the FCC will send the papers for the written test to the examiner, which you will do in his or her presence. The examiner, not you, will send them back to Gettysburg for scoring. If you did all your homework, the next piece of mail from that historic city will be a small envelope with your ham ticket in it. If you flunk the written examination you must wait another 30 days for a second crack at the test.

THE HIGHER LICENSES

To obtain a Technician, General, Advanced, or Amateur Extra Class license, you must appear in person at any district office or examination point of the FCC and take the test in person. You can go to any one of them, not necessarily the one nearest to your home. The Commission may permit the examinations for the higher class licenses to be administered at a location other than a Commission examination point by an examiner chosen by the Commission when it is shown by physician's certification that the applicant is unable to appear at a regular commission examination point because of a disability which prevents travel.

FCC Form 610
January 1976

Form Approved
OMB No. 52-R0069
Changes approved by
GAO-B-180227(R0036)

United States of America
Federal Communications Commission
Washington, D.C. 20554

APPLICATION FOR INDIVIDUAL AMATEUR RADIO STATION AND/OR OPERATOR LICENSE

ATTACH HERE the original or photocopy of your primary station license and an original or photocopy of any additional station license(s) as required by instruction F. If a license has been lost, mutilated, or destroyed, the circumstances must be explained in this block.

FCC USE ONLY (*Examiner's Report*)

ELEMENT	1(A) SPEED	1(B) SPEED	1(C) SPEED	2	3	4(A)	4(B)
PASSED							
FAILED							

DATE OF EXAM | FCC EXAMINER | DISTRICT NO.

OPERATOR CLASS FOR WHICH QUALIFIED

FCC USE ONLY (*Fee Stamp*)

CODE	NO	YES		EXAM	NO	YES		SCREEN	NO	YES		SIGN	NO	YES

I. READ ACCOMPANYING INSTRUCTIONS BEFORE FILLING OUT FORM. USE TYPEWRITER OR PRINT CLEARLY.

1. Licensed Amateur Radio Operators give the following information

PRIMARY STATION CALL SIGN | OPERATOR CLASS | EXPIRATION DATE

2. LAST NAME | FIRST NAME | MIDDLE INITIAL | 3. DATE OF BIRTH (*Month, Day, Year*)

4. MAILING ADDRESS (*all applicants must complete both items 4 and 5*)

NUMBER AND STREET

CITY | STATE

COUNTY | ZIP CODE

5. STATION LOCATION FOR FIXED OPERATION (*You cannot use a Post Office Box No. or General Delivery Address*)

NUMBER AND STREET (*or other indication of location*)

CITY | STATE

COUNTY

ITEMS 6A-13. Check the purposes for which you are filing this application and answer any associated questions. For example, if you check 6A you must also answer 6C.

6A. ☐ NEW OPERATOR AND PRIMARY STATION LICENSE

6B. ☐ CHANGE OF CLASS OF OPERATOR LICENSE

6C. OPERATOR CLASS APPLYING FOR (*check one box if you checked 6A or 6B*) ☐ NOVICE ☐ TECHNICIAN ☐ GENERAL ☐ AMATEUR EXTRA ☐ ADVANCED

7. ☐ RENEWAL ONLY | 8. ☐ RENEWAL AND MODIFICATION | 9. ☐ MODIFICATION ONLY | 10. ☐ NEW ADDITIONAL STATION

11. TYPE OF STATION BEING RENEWED/MODIFIED/OR ADDED (*See instruction 11.*)

☐ PRIMARY ☐ SECONDARY ☐ CONTROL ☐ AUXILIARY LINK ☐ REPEATER ☐ SPACE STATION

12. TYPE(S) OF MODIFICATION

☐ CHANGE OF STATION LOCATION ☐ OTHER (*specify*)

13A. ☐ REQUEST FOR SPECIAL CALL SIGN (*see instruction 13A*)

13B. SPECIAL CALL SIGN (*If previously held, give date it expired*)

14. Will the antenna structure exceed an overall height of 200 feet above ground level or one foot above the established airport elevation for each 100 feet of distance, or fraction thereof from the nearest boundary of such airport? If yes, complete and attach FCC Form 714, which is obtainable at any Commission office. ☐ YES ☐ NO

15. Would a Commission grant of your application be a major action as defined by Section 1.1305 of the Commission's Rules? See instruction 15 for explanation. If you answer yes, submit the statement as required by Section 1.1311. ☐ YES ☐ NO

16. Will the station you are applying for be remotely controlled? (See instruction 16.) ☐ YES ☐ NO

(OVER)

FCC Form 610 is the first step toward

The examinations for the various classes of amateur operator licenses are comprised of two or more of the following elements.

Element 1(A): Beginner's code test at five words per minute.
Element 1(B): General code test at 13 words per minute.
Element 1(C): Expert's code test at 20 words per minute.
Element 2: Written examination on rules and regulations essen-

17A. Do you have any other amateur radio application on file with the Commission that has not been acted upon? If yes, answer items 17B and 17C. YES ☐ NO ☐	17B. Purpose of other application	17C. Date Submitted
18A. Have you failed an amateur examination within the last 30 days? If yes, answer items 18B and 18C. YES ☐ NO ☐	18B. Class of Examination	18C. Date of Exam

19. City and State where you will take the amateur examination (*See instruction 19*)

20. Have you been convicted in a Federal, State, or local court of any crime for which the penalty imposed was a fine of $500 or more or an imprisonment of six months or more within 10 years previous to the date of this application? (*If "yes", see instruction 20*). ☐ YES ☐ NO

CERTIFICATION

I CERTIFY that all statements herein and attachments herewith are true, complete and correct to the best of my knowledge and belief and are made in good faith; that I am not a foreign government or a representative thereof; that I waive any claim to the use of any particular frequency or of the ether as against the regulatory power of the United States because of the previous use of the same whether by license or otherwise and that the station to be licensed, if any, will be inaccessible to unauthorized persons.

WILLFUL FALSE STATEMENTS MADE ON THIS FORM OR ATTACHMENTS ARE PUNISHABLE BY FINE AND IMPRISONMENT. U.S. CODE TITLE 18, SECTION 1001	21. SIGNATURE OF APPLICANT (*Do not print*)	22. DATE SIGNED

II. REQUEST FOR EXAMINATION PAPERS BY VOLUNTEER EXAMINER (*Applies only to Novice Class examinations by mail*)

1. NAME OF VOLUNTEER EXAMINER	2. NAME OF APPLICANT (*must be same as item 2 on front side*)

3. MAILING ADDRESS OF VOLUNTEER EXAMINER (*Number, Street, City, State and Zip Code*)

4. Check the operator license you have which allows you to administer the examination.

☐ GENERAL CLASS ☐ ADVANCED CLASS ☐ AMATEUR EXTRA CLASS	5. EXPIRATION DATE OF LICENSE
6A. Do you have a renewal application on file with us? If yes, answer item 6B. YES ☐ NO ☐	6B. DATE YOU FILED RENEWAL

CERTIFICATION OF VOLUNTEER EXAMINER

I certify that I am a licensed amateur radio operator, General Class or above; that I am at least 21 years of age; and that: (*Check appropriate box*)

☐ I have examined the above named applicant within the last ten days and he has successfully completed the prescribed telegraphy examination element set forth in Section 97.21 of the Commission's Rules.

☐ I have not examined the applicant in the International Morse Code since the applicant claims code test credit. Applicant's statement supporting code test credit is attached.

7. SIGNATURE OF VOLUNTEER EXAMINER (*Do not print*)	8. DATE SIGNED

III. PHYSICIAN'S CERTIFICATION
(*Required only if applying for an examination based on a protracted disability*)

☐ I certify that I have examined the applicant within the last 30 days and because of the protracted disability described below found him/her unable to travel to the nearest Commission examining point located at:

City and State

1. BRIEFLY DESCRIBE WHY APPLICANT CANNOT TRAVEL TO THE SPECIFIED NEAREST EXAMINING POINT.

2. NAME, ADDRESS, AND TELEPHONE NUMBER OF PHYSICIAN

3. SIGNATURE OF PHYSICIAN (*Do not print*)	4. DATE SIGNED

obtaining your amateur radio license.

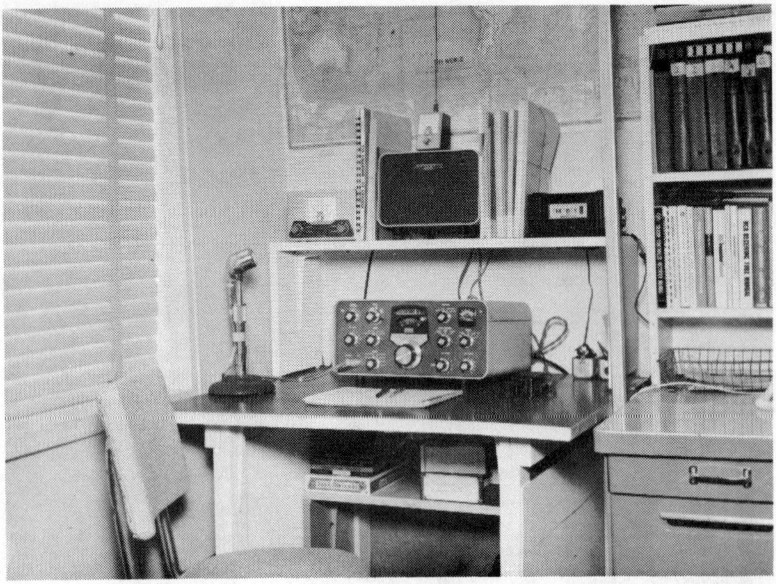

A corner only three-feet wide makes a cozy ham station for a New York apartment dweller. Shelf above transceiver holds accessories.

In his retirement home in Florida, Moe Segal, K4IVA, enjoys the comfort of a desktop full of ssb and 2-meter fm equipment.

Notice how the back edge of the "command console" at WA8TKC is angled downward to put the panels of the Drake 4-Line station in the line of sight of operator Ronald Wysong; a very good arrangement. Also note the huge map of the world covering the entire wall. Equipment, right to left: MS-4 speaker and ac power supply for transmitter; R-4B receiver; T-4XB transmitter; MN-2000 antenna tuner; L-4B kilowatt amplifier.

tial to beginner's operation, including sufficient elementary radio theory for the understanding of these rules.

Element 3: Written examination on general amateur practice and regulations involving radio operation and apparatus and provisions of treaties, statutes, and rules affecting amateur station and operators.

Element 4(A): Written examination on amateur practice involving intermediate level radio theory and operation as applicable to modern amateur techniques.

Element 4(B): Written examination on amateur practice involving advanced radio theory and operation as applicable to modern amateur techniques including the transmission of energy for measurements and observations applied to propagation, for the radio control of remote objects, and for similar experimental purposes.

Applicants for amateur radio operators licenses are required to pass the following examination elements.

Novice Class: Elements 1(A) and 2.

Technician Class: Elements 1(A), 2, and 3.
General Class: Elements 1(B), 2, and 3.
Advanced Class: Elements 1(B), 2, 3, and 4(A).
Amateur Extra Class: Elements 1(C), 2, 3, 4(A), and 4(B).

As stated previously, the Novice Class examination is not given by FCC personnel. It is given only by a volunteer examiner having the required qualifications. If you already have your Novice license, you only have to pass Element 3 in order to obtain a Technician Class license. Since you have already passed the Novice exam, you will be given credit for Elements 1(A) and 2. All you

Boys not even in their teens and men old enough to be their grandfathers share the hame hobby with equal enthusiasm. William G. Welsh, W1SAD/6, thrills a young friend by letting him participate in an overseas conversation.

have to take now is another written exam which is definitely more extensive than the first one. If you have a General Class license and want to upgrade to Advance, you just take Element 4(A). To upgrade from General to Amateur Extra, you must take Elements 1(C), 4(A), and 4(B). Your test is scored immediately by an FCC examiner, but your actual license will be mailed to you from Gettysburg. If you flunk, you may return for another try after 30 days.

It is not necessary to obtain a Novice license before trying for a higher class license. In fact, many people qualify for a General or Advanced license on their first try. Of course, they must pass all of the elements required for the license class they wish to obtain.

When he was first licensed in 1922 as 6BPI, at the age of thirteen, this ham had much less equipment than he now enjoys as K7UGA. He dropped out of the game during the depression of the 1930s but never lost his interest. His amateur experience proved valuable during World War II, when he taught radio code and theory to Chinese aviation cadets. Again bitten by the bug in 1962, he quickly got into the swing of things and acquired the call K7UGA for his permanent home in the Southwest and K3UIG for his Washington base. He also operates mobile rigs in his car and boat. Does the face look familiar? It should, because it belongs to Barry Goldwater, senator from Arizona.

The new Rules and Regulations now permit you to obtain an Amateur Extra license without previously holding a lower grade license. However, don't underestimate the Amateur Extra: it takes a lot of preparation.

There is a little hidden bonus in the new regulations. If you pass either the 13- or 20-word-per-minute code test but fail the respective written exam, you are given a code credit certificate good for one year, providing you return to the issuing office.

Eugene M. Link, WØIA, a retired army colonel, has won public-service citations for organizing an amateur network devoted to weather reporting.

ASSIGNMENT OF CALL SIGNS

For the purposes of call-sign assignment, the United States is divided into ten districts, as follows:

First: All the New England states of Connecticut, Rhode Island, Massachusetts, New Hampshire, Vermont, and Maine.

Second: New York and New Jersey.

Third: Delaware, Maryland, Pennsylvania, and the District of Columbia.

Fourth: Alabama, Florida, Georgia, Kentucky. North Carolina, South Carolina, Tennessee, Virginia, and the territories of Puerto Rico and the Virgin Islands.

Fifth: Arkansas, Louisiana, Mississippi, New Mexico, Oklahoma, and Texas.

A nice wall decoration for the radio shack: a map of the United States with the boundaries of the ten call-letter districts outlined with a marking pen. Use colored pins to indicate the locations of stations worked.

Sixth: The states of California and Hawaii, and the Pacific islands not assigned to the Seventh District.

Seventh: Arizona, Idaho, Montana, Nevada, Oregon, Utah, Washington, Wyoming, Alaska and adjacent islands.

Eighth: Michigan, Ohio, and West Virginia.

Ninth: Illinois, Indiana, and Wisconsin.

Tenth: Colorado, Iowa, Kansas, Minnesota, Missouri, Nebraska, North Dakota, and South Dakota.

By international agreement, the ham stations of the world are identified by call signs having exclusive prefix letters, or in some cases number/letter combinations starting with the number. The United States prefixes are A, W, K, and N. All U.S. amateur call signs consist of one or two prefix letters, the first of which is any of the four just mentioned; then a number from 1 to Ø for the district; and finally one, two, or three letters. The tenth district is always spoken as "zero," never "oh" or "naught."

Over the air, you will hear frequent references at call signs as "one-by-three" or "two-by-two," etc. This is a short way of describing the calls. The word "by" is substituted for the district

If you can't tell from his wild shirt, you can surely tell from his license plate that Ralph E. Thomas and his XYL (ham for wife) live in Hawaii.

number. The numbers before and after the word "by" refer to the number of letters before and after the district number. Some examples include:

W2UN—one-by-two
AA4X—two-by-one
KB6FL—two-by-two
N5ABC—one-by-three
KZ9DLM—two-by-three

There are no one-by-one or three-by-three combinations.

Certain two-letter prefixes are reserved for stations outside the 48 contiguous states. Examples: KP for Puerto Rico, KH for Hawaii. By special permission of the government, U.S. personnel living in Japan may have FCC licenses with the prefix KA.

For the first time in amateur history, specific groups of calls are being assigned according to the class of the applicant's license. The higher the class, the more distinctive the call. These calls are grouped according to license class in the following manner.

Group A for Amateur Extras consists of one-by-twos, two-by-ones, and two-by-twos.

Group B for Advanced operators consists of two-by-twos.

Group C for Generals and Technicians consists of one-by-threes.

Group D for Novices consists of two-by-threes.

It is interesting to note here that the prefix N was the exclusive property of the U.S. Navy for as far back as it has used radio, and was released to the amateur radio fraternity only in 1977.

Amateur prefixes that start with a number instead of a letter are generally heard from the smaller countries of the world. Don't be surprised to hear calls like 3A2AH in Monaco, 3V8AL in Tunisia, 4X4NUD in Israel, or 9V1NR in Singapore.

LOCATION OF FCC OFFICES

The FCC has 28 permanent district offices where tests are held almost daily during normal business hours. It also has about 50 *examination points* where tests are given at various times of the year. Here are the addresses and phone numbers of the offices. Before you visit one, by all means invest in a telephone call and inquire about its hours. If the nearest office is quite a distance from your home, write to it for the location and schedule of an examination point that might be closer.

You are not required to take the exam only at an FCC facility in your call-sign area. Many applicants arrange vacation trips to resort areas that happen to have FCC offices, and they take the tests at their leisure. How about the Miami office in the winter, or the one in Alaska in the summer? Your test papers wind up in Gettysburg, regardless.

Addresses of FCC Field Offices

The addresses and telephone numbers of the FCC field offices alphabetically by state, including the field offices in Puerto Rico and the District of Columbia are as follows:

ALASKA, Anchorage
U.S. Post Office Building Room G63
4th & F Street, P.O. Box 644
Anchorage, Alaska 99510
Phone: (907) 265-5201

CALIFORNIA, Long Beach
3711 Long Beach Blvd.
Suite 501
Long Beach, California 90807
Phone: (213) 426-7955

CALIFORNIA, San Diego
Fox Theatre Building
1245 Seventh Avenue
San Diego, California 92101
Phone: (714) 293-5460

CALIFORNIA, San Francisco
323A Customhouse
555 Battery Street
San Francisco, California 94111
Phone: (415) 556-7700

COLORADO, Denver
Suite 2925, The Executive Tower
1405 Curtis Street
Denver, Colorado 80202
Phone: (303) 837-4053

DISTRICT OF COLUMBIA, Washington, D.C.
1919 M Street N.W. Room 411
Washington, D.C. 20554
Phone: (202) 632-8834

FLORIDA, Miami
51 S. W. First Avenue, Room 919
Miami, Florida 33130
Phone: (305) 350-5541

FLORIDA, Tampa
809 Barnett Office Building
1000 Ashley Drive
Tampa, Florida 33602
Phone: (813) 228-2605

GEORGIA, Atlanta
440 Massell Building
1365 Peachtree St. N.E.
Atlanta, Georgia 30309
Phone: (404) 881-7381

GEORGIA, Savannah
238 Federal Office Bldg. and Courthouse
125 Bull Street, P.O. Box 8004
Savannah, Georgia 31402
Phone: (912) 232-4321 ext. 320

HAWAII, Honolulu
502 Federal Building
P.O. Box 1021
355 Merchant Street
Honolulu, Hawaii 96808
Phone: (808) 546-5640

ILLINOIS, Chicago
230 South Dearborn Street, Room 3935
Chicago, Illinois 60604
Phone: (312) 353-0195

LOUISIANA, New Orleans
829 F. Edward Hebert Federal Bldg.
600 South Street
New Orleans, Louisiana 70130
Phone: (504) 589-2094

MARYLAND, Baltimore
819 Federal Building
31 Hopkins Plaza
Baltimore, Maryland 21201
Phone: (301) 962-2727

MASSACHUSETTS, Boston
1600 Customhouse
165 State Street
Boston, Massachusetts 02109
Phone: (617) 223-6608

MICHIGAN, Detroit
1054 Federal Building
231 W. Lafayette Street
Detroit, Michigan 48226
Phone: (313) 226-6077

MINNESOTA, St. Paul
691 Federal Building & U.S. Courthouse
316 N. Robert Street
St. Paul, Minnesota 55101
Phone: (612) 725-7819

MISSOURI, Kansas City
1703 Federal Building
601 East 12th Street
Kansas City, Missouri 64106
Phone: (816) 374-5526

NEW YORK, Buffalo
1307 Federal Building
111 W. Huron Street
Buffalo, New York 14202
Phone: (716) 842-3216

NEW YORK, New York
201 Varick Street
New York, New York 10014
Phone: (212) 620-3435

OREGON, Portland
1782 Federal Office Building
1220 S.W. 3rd Ave.
Portland, Oregon 97204
Phone: (503) 221-3097

PENNSYLVANIA, Philadelphia
11425 James A. Byrne Federal Courthouse
601 Market Street
Philadelphia, Pennsylvania 19106
Phone: (215) 597-4410

PUERTO RICO, San Juan
747 Federal Building
Hato Rey, Puerto Rico 00918
Phones: (809) 753-4008 or 4567

TEXAS, Beaumont
323 Federal Building
300 Willow Street
Beaumont, Texas 77701
Phone: (713) 838-0271

TEXAS, Dallas
Earle Cabell Federal Bldg. &
U.S. Courthouse
13E7, 1100 Commerce Street
Dallas, Texas 75242
Phone: (214) 749-3243

TEXAS, Houston
5636 New Federal Office Bldg.
515 Rusk Avenue
Houston, Texas 77002
Phone: (713) 226-4306

VIRGINIA, Norfolk
Military Circle
870 North Military Highway
Norfolk, Virginia 23502
Phone: (804) 461-4000

WASHINGTON, Seattle
3256 Federal Building
915 Second Ave.
Seattle, Washington 98174
Phone: (206) 442-7610

CALL PLATES FOR YOUR CAR

In recognition of the extraordinary public services rendered by hams during periods of disaster and emergency, many states permit them to obtain license plates for their cars that duplicate their call signs. Of course, this is a highly prized privilege—other "low

If you don't want to bother with changing the registration of your car, you can still identify yourself as a ham by mounting a cast aluminum call plaque under the state license tag. Available in a variety of sizes and styles, these plates can be ordered from most amateur supply firms.

number" and initial plates are not nearly so distinctive. Check with the motor vehicle licensing authorities of your own state. Where the plates are available, you only have to show your ham ticket to obtain them. In most cases there is a small extra charge, inasmuch as the plates are made to order.

If you are lucky enough to get call-sign plates, you can expect to be asked a lot of questions by curious passers-by, and also you

Many radio clubs offer would-be hams free instructions that enable them to qualify for licenses. Clubs that have stations also give students the opportunity to observe and even participate in live communications with modern equipment. This photo shows the highly popular rig at WB2JSM, the station of the Hall of Science Radio Club in a former building of the 1939 New York World's Fair. At the mike is one of the club's instructors, Gary Kerness, WB2CWK, who is responsible for getting many happy operators on the air with their own stations.

will probably make the acquaintance of a lot of other hams. On the road, you'll hear many greetings blasted out on car horns in International Morse, so you'd better keep your code up. Many hams have installed telegraph keys on the steering columns of their cars, wired across the horn switch contacts, so that they can send properly. It's all a lot of fun.

6

Going on the Air

TYPES OF TRANSMITTERS

The continuous-wave transmitter is the simplest of all. It generates a sharp single-frequency signal, which is controlled by a hand key to form the dits and dahs of the International Morse Code. Technically, this form of communication is *radiotelegraphy,* but universally it is called simply CW.

There are no longer any factory-built CW-only transmitters on the amateur market, but CW capability is included in virtually all of the more advanced types of amateur transmitters.

This is the busy front end of the Atlas Model 350-XL, a six-band, 100% solid-state transceiver with enough controls to keep any knob twister happy. Digital readout and plug-in auxiliary vfo are optional. The transmitter is rated at 350 watts.

Surprisingly effective low-power CW transmitters can be assembled with parts salvaged from discarded radio and television receivers. The ham radio magazines frequently publish articles on the subject.

When voice facilities are added, a CW generator becomes a "phone" transmitter. Voice signals superimposed on the narrow CW carrier cause it to change in amplitude and also to broaden out both above and below its basic frequency. For example, if an operator's speech varies up to 2000 hertz, a 14.3-MHz carrier widens to 14.302 MHz at one extreme and to 14.298 MHz at the other. The two variations from the carrier frequency are called *sidebands*. This general method of operation is called *amplitude modulation* or am, and is used by all standard broadcasting stations in the 550 to 1600 kilohertz band; also by some stations in the aviation, weather, and public-safety services. Most am equipment is relatively easy to build, adjust, and maintain.

The other general method of operation is called *frequency modulation* or *fm*. In this type of modulation, the amplitude of

The designers of the Ten-Tec Triton IV have produced a piece of ham gear that fits perfectly in a panelled den or any other room. This is a total solid-state transceiver of advanced construction. The mike is shown at the left and electronic keyer at the right.

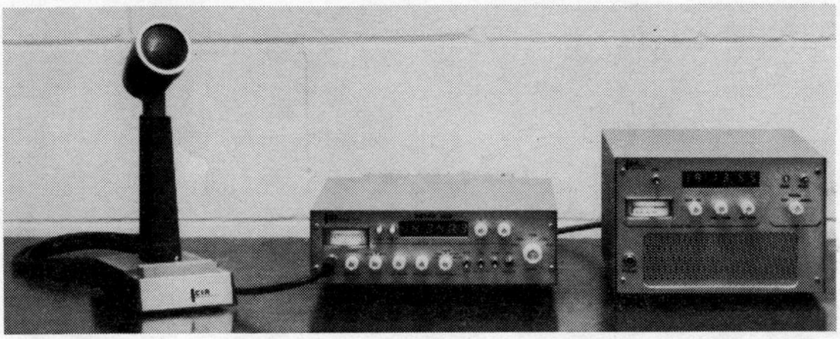

The compact Astra 200 solid-state transceiver in the center is flanked on the right by an operating cosole that contains a 6-digit, 24-hour clock; a phone patch; a loudspeaker; and a 10-minute timer with both aural and visual indication. This is a very cozy little communications center.

the carrier remains constant and the frequency is changed by the superimposed voice signals. Although fm is used mainly for high-fidelity broadcasting, it is also advantageous for relatively short-distance mobile communication because it is less susceptible to noise interference than am. In the absence of signals a properly adjusted fm receiver appears to be absolutely dead, but it comes to life loud and clear when they do arrive. This trick is accomplished by "limiting"—literally desensitizing—the receiving circuit so that it does not respond at all to weak signal below a critical value but is "captured" readily by strong ones above that level. In amateur practice, fm is used almost entirely on the 144-, 220- and 440-MHz bands. (See Chapter 12.)

Single-Sideband Operation

An am signal occupies a rather wide bandwidth. It can be cut almost in half, however, by the simple expedient of eliminating either of the sidebands. Intelligibility is not affected in the slightest since the two sidebands carry absolutely identical voice information. It is also possible to squeeze out the carrier, either partially or fully, so the method is called *single-sideband suppressed carrier*, or, more generally, just *single-sideband*, or *ssb*. What is gained from all this? Two very important advantages over am are:

(1) The transmitted signal occupies much less space on the air. In fact, it is entirely practical for two pairs of stations using the same carrier frequency but different sidebands to carry on conversations with little or no mutual interference. In effect, ssb doubles the number of stations that can be accommodated within the limits of any band.

(2) From the standpoint of hams living in crowded areas served by several television stations, an extremely important feature of

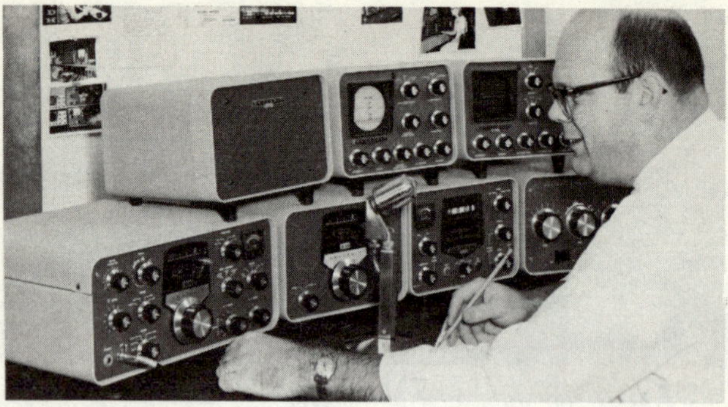

This very professional looking amateur station is strictly home assembled, made from Heathkits. The happy operator is Alan Robertson, K8BLL, Benton Harbor, Michigan. Left to right, on table: SB-102 five-band transceiver, SB-640 external linear master oscillator, SB-630 station console, SB-200 kilowatt amplifier. Top: SB-600 speaker, SB-620 signal analyzer, SB-610 monitor scope. Never a dull moment in the life of this DX worker!

ssb is its virtual freedom from "tvi"—the tendency to create interference in tv receivers. Most am transmitters are very bad in this respect because they radiate many spurious signals (harmonics) that are multiples of the widely varying carrier and its two sidebands. These show up on the picture tube as herringbone patterns or in the sound as muffled but understandable voice. Extensive shielding of the transmitter and filtering of its circuits are usually necessary to alleviate the trouble. With ssb, however, one sideband and all or most of the carrier are eliminated within the transmitter itself, and it is a comparatively simple job to clean up the remaining signal that is transmitted.

Single sideband was adopted for commercial radiotelephony in 1927, but for a long time thereafter it had very little amateur

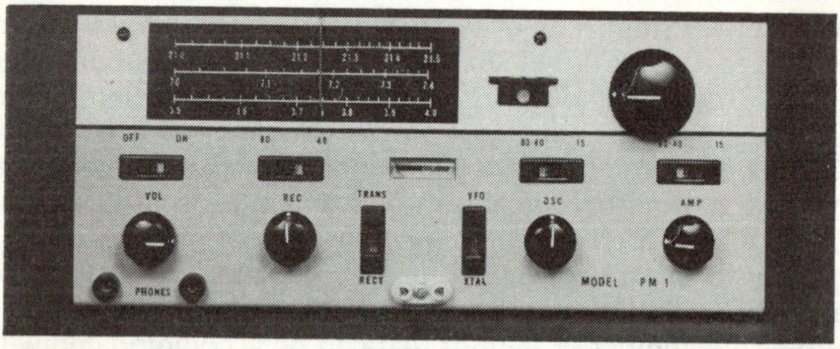

The Ten-Tec "PM" series low-power CW transceivers use modular construction, are reliable because they are relatively simple, and are especially appealing to beginners because of their low prices. The model shown coveres 80, 40, and 15 meters. The only power supply needed is an inexpensive lantern battery.

You begin to appreciate the international flavor of ham radio when you work your first foreigner, perhaps SM5CLW who lives in Sweden. Lajos Nemeth not only speaks excellent English but also operates a whole tableful of U.S.-made Drake equipment.

application because of its complexity. Early single-sideband ham equipment was generally unstable, required very critical tuning, and produced peculiary distorted signals very much like the voice of Donald Duck. These shortcomings disappeared with the advent of new techniques and improved circuit components. Scores of manufacturers now offer a wide variety of dependable and completely satisfactory ssb gear.

It should be pointed out that ssb transmitters are amplitude modulated, and therefore should be considered am equipment. However, in radio language am specifically means conventional double-sideband modulation, as distinguished from ssb.

The superiority of ssb is so evident that am has pretty well disappeared from the ham bands; in fact, it is unusual to find ssb sets that offer optional am transmission.

There is an arbitrary and unofficial operating practice among some hams to use only upper sideband on 10, 15, and 20 meters and only lower sideband on 40 and 80 meters. There are even a few ssb rigs in which no other choice is possible. These restrictions are senseless. Many times a conversation will start on one sideband and then become enmeshed in interference, but a quick switchover to the other sideband might readily clear the air.

TRANSCEIVER OR TRANSMITTER?

One you've passed your Novice test and know for sure that your license is on the way, you will start to think seriously about going on the air. What type of equipment should you consider?

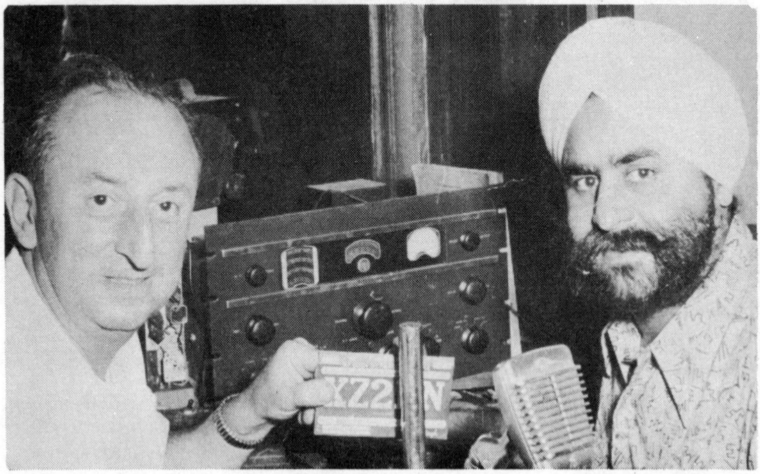

David Marks, W2APF, Albany, N.Y. is the ham fraternity's roving ambassador. First he works amateurs in odd corners of the world, and then takes off to visit them in person. Known everywhere as "Uncle Dave," he is shown here with XZ2KN, Tara Singh, of Rangoon, Burma. He has made more personal contacts of this kind than any other ham. His home "shack" is filled with souvenirs of his travels, many of them both exotic and expensive.

This is a very desirable feature in transmitters and transceivers of the hybrid type (a mix of tubes and solid-state devices). The small, quiet fan blows air through the chassis. By preventing components from overheating, it greatly lengthens their lives.

As a Novice you are limited to CW on a few narrow slices of the 80-, 40-, 15-, and 10-meter bands, but you certainly won't stay a Novice for long. When you qualify for higher licenses your frequency allowances will be very wide, so you might as well prepare for them from the start with the right gear.

In Chapter 4, which deals primarily with receivers, mention was made of the *transceiver* which is a combination transmitter and receiver. This is by far the most popular type of ham equipment. Some models are entirely solid-state; others are solid-state in the receive mode and use a mix of solid-state devices and a couple of tubes in the transmit mode.

Features shared by most transceivers include the following: choice of phone or CW; choice of sideband; break-in CW keying; quick switching from receive to transmit, either manually by a button on the microphone used for phone or automatically by VOX, which stands for *voice operated transmitter;* tuning meter; frequency calibrator or direct digital readout of frequencies; coverage of as many as 10 or 11 band segments; noise blanker; cooling fan; and connections for outboard accessories.

In a basic transceiver the transmit and receive frequencies are identical. With a transmitter/receiver duo, the frequencies can be different. This flexibility can be advantageous at times. If you ever feel the need for it, you can easily add an external vfo, which is short for *variable frequency oscillator.*

The Swan Model 270 ssb transceiver is a complete station in one package, needing only a power source and an antenna. With its self-contained ac/dc power supply, it is equally suitable for fixed and mobile use.

Direct frequency readout from both the Yaesu FT-101 and the Collins "S" line transceiver is available from the Spectronic Digital Display, shown here on top of the FT-101. A simple connection to the existing set is all that is needed.

A representative collection of transmitting equipment is pictured here for you to drool over. No prices are given, as they are too variable. Keep up with the ham market by studying the advertisements in the ham magazines.

TRANSMITTER POWER

Although the distance over which signals from a short-wave station can be received depends more on atmospheric conditions than on transmitter power, a good start from the latter is helpful nevertheless. By FCC regulations, hams other than Novices are permitted to use 1000 watts of power (a *gallon*!) as measured at the final amplifier stage of the transmitter. With this figure considered high power, anything below about 50 watts is low power, and up to about 250 watts is medium power. Actually, there is no such thing as a self-contained one-kilowatt ham transmitter.

Most transceivers designed for the 10 through 80 meter bands are rated at about 150 watts. If an owner wants more output, he adds an amplifier with the legal 1000-watt rating between the transceiver and the antenna. In this combination the transceiver is called an *exciter* and the amplifier is a *linear*. The two units are

The Heathkit HW-101 is a medium-price, medium-power transceiver much favored by do-it-yourselfers because it is rather simple to assemble and has a professional appearance.

cabled together and are controlled by a switch on the linear amplifier. When the switch is off, the exciter is connected directly to the antenna. When it is on, the exciter is connected to the input of the linear and the antenna is connected to its output.

An 1000-watt linear amplifier is a large, heavy, and rather costly piece of equipment. Some models might require connection to a 240-volt ac line rather than to a normal 120-volt ac circuit. You can get along perfectly well without a linear, but if you have space in your ham shack and money in the bank, you can always add one to your collection of gear.

Very Low-Power CW

An old game resurfaced in ham circles in the early 1970s. It's called "QRPing," from the international Q signal meaning

The first time you see this piece of equipment you might ask: "What is this, the control panel of a small space ship?" No, it's only a new ham transceiver, the Hy-Gain 3750, fitted with 36 controls that give the operator complete mastery of all control functions.

The boxing glove on top of the Swan 700CX is meant to show that this transceiver packs a real punch—700 watts worth. It covers the 10-, 15-, 20-, 40-, and 80-meter bands on ssb, am, and CW. A high-quality station in one package.

"decrease power." The general idea is to see how much DX can be accomplished with little power. Normal transmitters run all the way to 1000 watts, but the adventurous QRP boys consider anything more than 2 or 3 watts excessive. Equipment of this rating is usually referred to as "flea power" or "peanut whistle."

Manufacturers were quick to recognize a trend. Ten-Tec offers several factory-built low-power CW transceivers. The Heathkit HW-8, a do-it-yourself job, is shown on page 97. The circuitry of

The handle on the end of the Henry Radio "Tempo One" transceiver means that the unit is compact enough to be portable. This is a full ssb set, with choice of an ac power pack for fixed-station use and a dc power pack for mobile use.

this little set is fairly advanced. Portable operation on a couple of lantern batteries is practical.

TRANSMITTER ACCESSORIES

You have to read the fine print of some catalogs very carefully to learn that many expensive transceivers do not include a power supply of any kind. Both ac and 12-volt dc power supplies are usually available as accessories, the latter being intended primarily for use in cars and other vehicles. Either type can usually be accommodated inside the loudspeaker enclosure sold by the transceiver manufacturer as a matching unit.

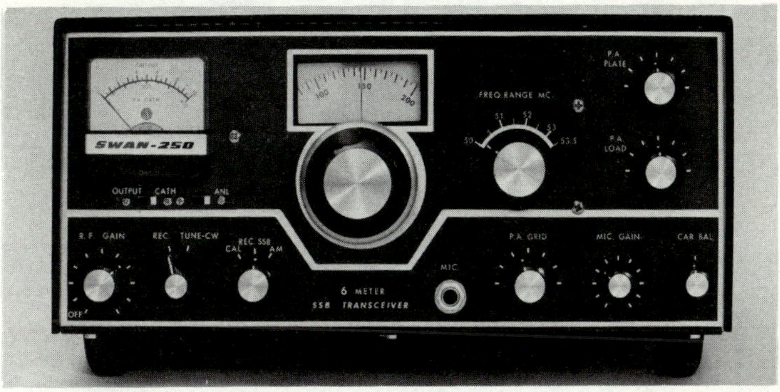

The advantages of ssb are extended to the 6-meter band in the Swan 250, which is devoted exclusively to coverage of 50 to 54 megahertz. It is particularly effective for mobile use because a small antenna serves well. An external speaker and ac or battery power pack are required.

You can easily fill a table with other accessories, such as a phone patch, an antenna bridge, a 24-hour clock, a field-strength meter, a station monitor using a cathode-ray oscilloscope, an electronic keyer, a recorder for taping interesting conversations, an illuminated globe, and similar goodies.

A loudspeaker is self-contained in some multiband ssb transceivers, but it is generally very small and squawks rather than speaks. The manufacturers often suggest the use of a larger external speaker. Absolutely no sets of this type include either a key or a mike. These are rather personal items, and the choice of styles and prices is very wide. Since there is nothing in a key or mike to wear out, you should buy good ones at the beginning of your ham career.

About the only ham gear that includes both a microphone and a loudspeaker are 2-meter and 6-meter transceivers. In some hand-held models the speaker doubles as the mike, as in most intercom systems. In the larger models the speaker is still inside the set, but the mike is separate and often connected to the transceiver by a coiled cord. (See Chapter 12).

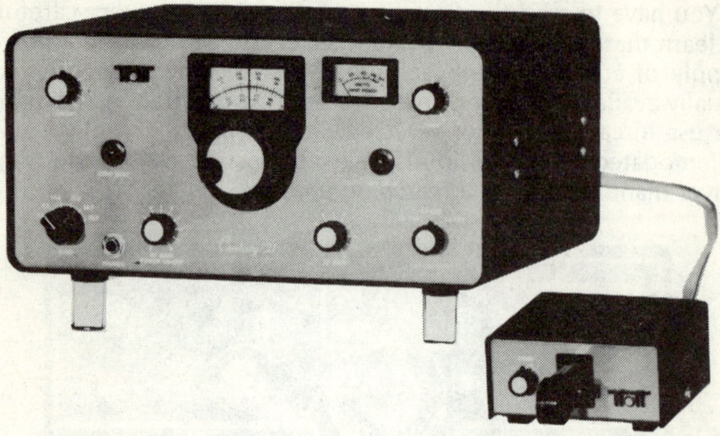

This is the Ten-Tec Century 21, a special transceiver intended for CW buffs who'd rather use the International Morse Code and a key than gab into a microphone. It is all solid-state, runs at 70 watts, and features instantaneous break-in and coverage of four bands. Electronic keyer, at right, is an optional accessory.

The straight hand key hasn't changed in design or appearance since the Civil War. Because it's simple and easy to use, you should start with this type. Later, you may want to try a semiautomatic key known variously as the "speed key," "side swiper," or "bug." The second term is derived from the side-to-side motion of the operator's fingers, as distinguished from the up-and-down tapping motion used with a straight key. The third term is derived from the appearance of the trademark of the original semiautomatic key, which bore the trade name *Vibroplex*.

A speed key is *semi*automatic. When you press with your thumb against the left side of the protruding handle you set a short metal reed into vibration. Fitted with contact points, this produces a series of dits at a speed determined by the position of a sliding weight. The vibration and the dits continue for several seconds if you keep your thumb against the handle. Actually, only a very short pressure is needed, since the longest dit character is the number 5, with five dits. To make dahs, you relax your thumb and *tap* the other side of the handle with your first finger. To make dit-and-dah characters, you wiggle your fingers back and forth. This technique is slightly tricky and takes practice. However, learning is not too difficult because half of it is the same as straight-key operation. Twenty-five or thirty words per minute with a straight key is hard work, but is easy with a bug.

Hams who consider themselves fancy operators often use automatic electronic keyers. These are essentially adjustable double oscillators; one section produces perfect dits, the other perfect

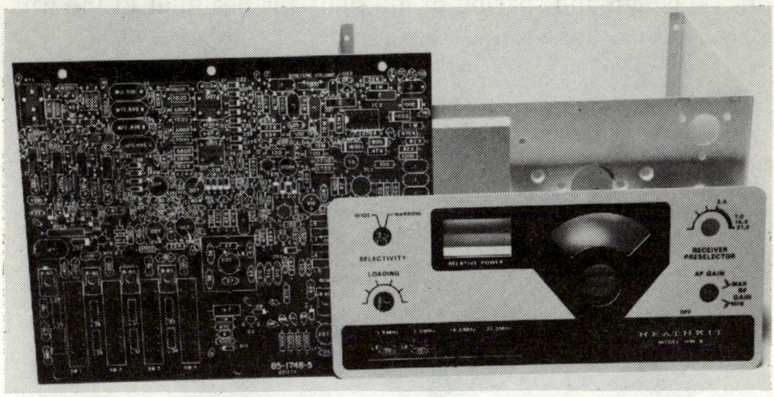

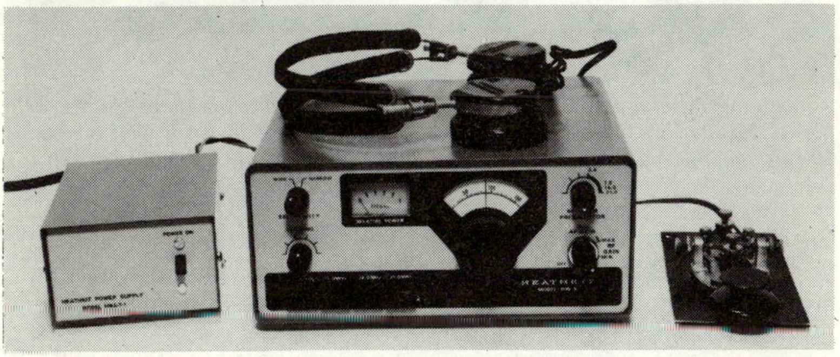

The Heathkit HW-8 is a low-power CW transceiver available in kit form. The top photo shows some of the kit parts. The bottom photo shows a complete station with ac power supply, headset, and key. This transceiver uses 98 resistors, 260 capacitors, 46 solid-state devices and numerous other components. Note the busy appearance of the empty circuit board. Operating this peanut whistle is fun.

dahs, at speeds up to 60 wpm (words-per-minute), which is faster than most hams can copy. The manual control is a paddle key that looks very much like a bug but is quite different in that it has identical left and right finger positions which make only momentary contacts, like a hand key or the dah side of the bug. Press it with the left thumb, and the electronic unit spews out dits; press it with the first finger, and it spews out dahs.

Learning to use an electronic keyer, after you have mastered either the straight key or the bug, is like trying to talk Italian after you have spent four years studying French. You can expect to become thoroughly balled up, and you won't be able to produce intelligible Morse until you've devoted considerable time to retraining yourself. The effort is worth the trouble, because the precise length and spacing of the signals is beautiful to the ears.

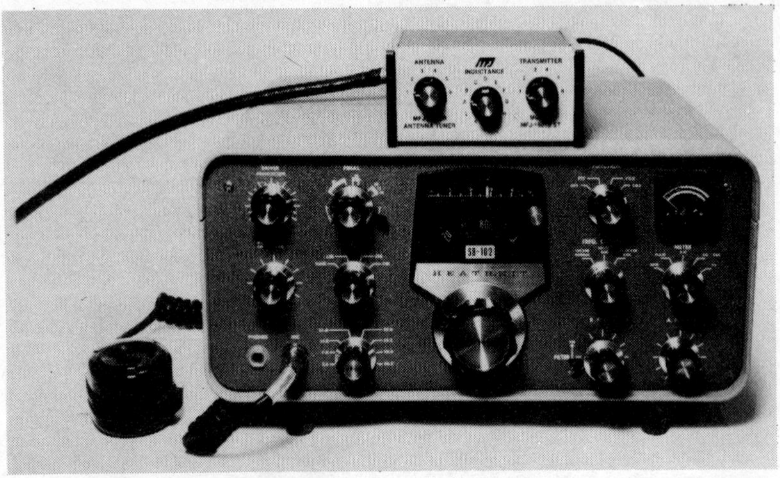

An extremely useful accessory in the ham shack is an antenna tuner like the MJF Model 16010. This tuner permits almost any transceiver to work well on bands 10 through 160 meters with random lengths of almost any kind of wire. This is great for experimentation.

Of course, this advantage isn't obtained for nothing; a perfectly good straight key can be bought for less than a dollar and a side swiper for about $16, but an electronic keyer with paddle runs about $50 to $100.

It was pointed out in Chapter 4 that high-fidelity audio response is not only unnecessary in ham work but downright undesirable. Hams are limited to voice modulation and are specifically prohibited from transmitting music or any other form of entertainment. This means that you should look for a mike designed for

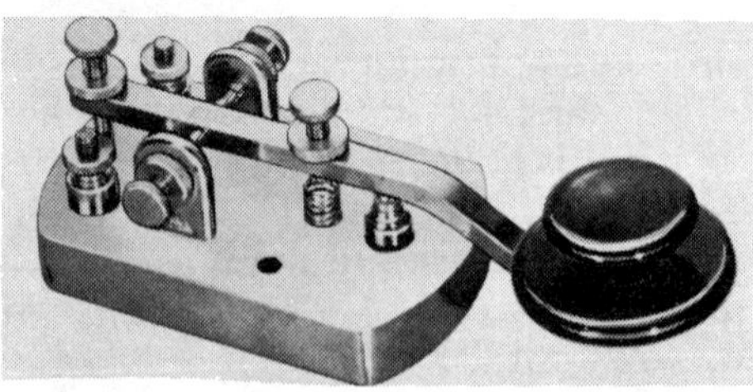

The Johnson Model 114-320 is a straight key of rugged construction.

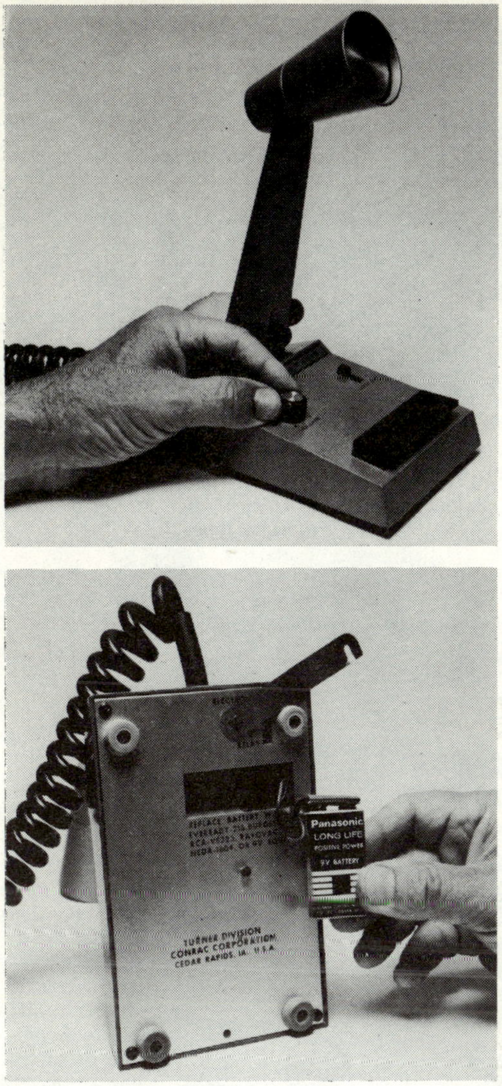

A relatively new development in microphones is the power type which adds considerable punch to many transceivers. Its secret is a built-in, battery-operated transistor amplifier with its own volume control. Representative model is the Turner "Plus Three."

communication purposes rather than for public-address or broadcasting service.

The two types favored by hams are the ceramic and the dynamic, and both are available in a wide variety of qualities and

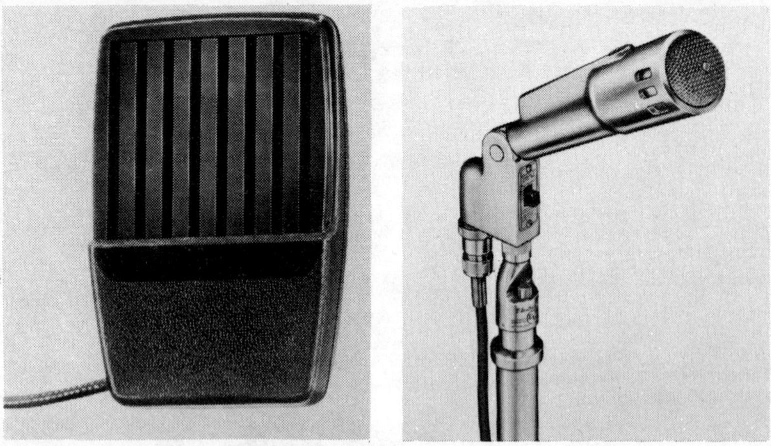

Representative microphones used by many hams. Left: Electro-Voice Model 721 is a low-cost ceramic type. Right: Electro-Voice Model 664 is a more rugged dynamic type.

The Heathkit "Cantenna" is a dummy antenna (actually a large resistor immersed in oil) used during adjustments on a transmitter when it is desired not to send out live signals. It is generally used with a coaxial switch (center object) so that the transmitter can be connected quickly to one of several actual antennas or to the dummy antenna. This is an invaluable arrangement for experimental work.

prices. A third type known as the carbon is universally used in telephone instruments, and was also found in many World War II field radio sets. Surplus military mikes of this model can be bought very cheaply. However, they are poor bargains because they require a source of low-voltage dc, which is not usually on tap in conventional ham transmitters. The ceramic and dynamic

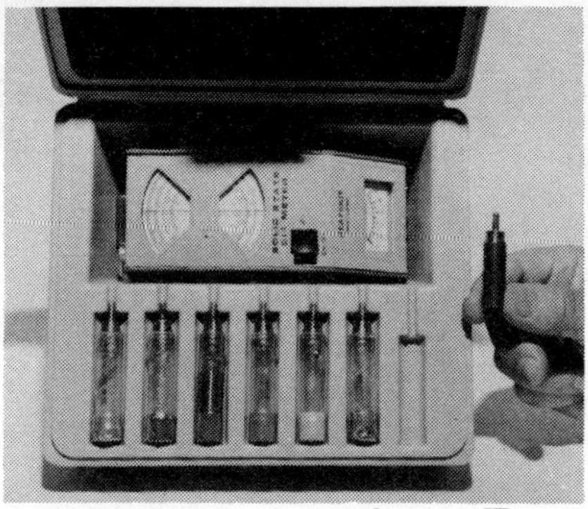

Above: Heathkit Model 1250 Dip Meter has plastic carrying case for outdoor use. Right: Coils for wide frequency spread plug into top edge of chassis. Dial is thumb operated. Below: Standard 9-volt transistor battery is self contained.

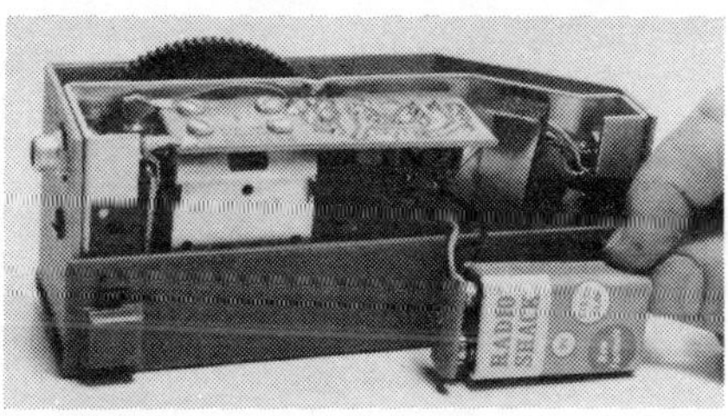

mikes generate their own current when spoken into, and are therefore much simpler to use.

A little accessory that is very useful for checking and adjusting transmitters and antennas is the "dip meter." This is calibrated in terms of frequency from 1.6 to 250 MHz. Containing two transistors and a handful of other parts and powered by a self-contained 9-volt battery, it is particularly valuable for work on antennas and mobile equipment.

In effect a dip meter is a combination receiver-transmitter. It can respond to signals from equipment or it can inject signals of its own into the latter, as conditions require.

The term "grid-dip meter" is often heard in ham operations, but it applies only to older forms that used vacuum tubes. The term "dip meter" specifically means a solid-state instrument. A typical one, made from a Heathkit, is shown on the preceding page.

7

The Antenna Radiates the Signals

Before a study of transmitting antennas is begun, a clear understanding of the terms frequency and wavelength is necessary. The two terms are different ways of describing the same alternating current, each being more convenient than the other for some purposes.

FREQUENCY VS. WAVELENGTH

As you probably know, alternating current is so called because it starts from zero, rises evenly to a maximum value in one direction, falls off to zero, then reverses its direction of flow and goes through exactly the same rise and fall. Each rise and fall is called an alternation, and two complete alternations comprise a cycle. It is common to refer to one alternation as *positive* and the other as *negative,* but this is slightly misleading because the word negative infers uselessness. Actually, the two alternations are absolutely identical in their ability to do work.

The rate at which complete cycles of ac are produced is called the *frequency*. Frequency is expressed as so many *hertz*. A hertz is one cycle per second. The high frequencies used for radio signals makes the use of multiplier prefixes desirable. For example, instead of 600,000 hertz, it is easier to write and say 600 kilohertz, or simply 600 kHz. Going still higher, 2,000,000 hertz becomes 2 megahertz, or 2 MHz.

This beautiful 50-foot Tristo tower is next to the garage of Carl Mosley, WØFQY, St. Louis, Missouri. How did he get that elaborate antenna in the air?

By pressing a button, that's how! Tower sections not only telescope, but the lowered assembly tilts over to permit easy and safe adjustments on the antenna proper, as Carl demonstrates. He is manufacturer of the popular line of ham antenna bearing his name. Tower is entirely self-supporting.

The lower the frequency of the ac, the longer the spacing or time interval between the ups and downs of current; the higher the frequency, the shorter the distance. Since we know fairly definitely that the speed of electricity is 300,000,000 meters per second, we have only to divide this figure by the frequency of a current to determine the separation between successive up or down points as impulses move along wires or away from a radiating body such as an antenna. This spacing is the *wavelength,* and is always expressed in the metric unit of length, the meter. Thus, a radio signal having a frequency of 30 MHz (30 million hertz) has a wavelength of 10 meters. This is equivalent to 32 feet, 8 inches in the English system.

A straight, simple wire radiates energy most effectively when it is a *half* wavelength long; in other words, it must be designed for a specific wavelength or frequency. Why a half wave and not a

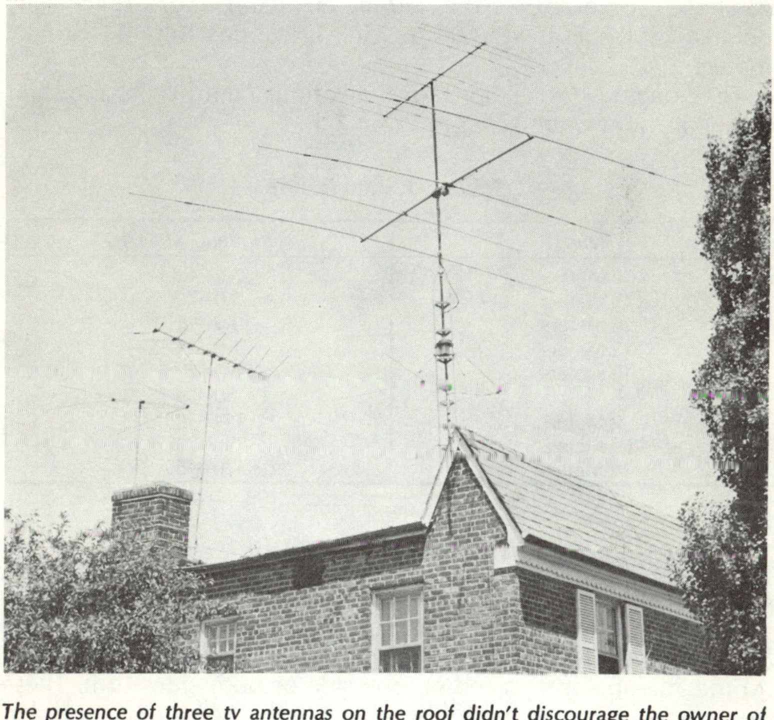

The presence of three tv antennas on the roof didn't discourage the owner of this house from adding a 15-foot tower topped off with an 8-element beam for 6 meters, and a 4-element beam for 10/15/20 meters. Beams are turned by rotator motor halfway up the tower. There is still no interference with television reception.

full wave? Because a half wave represents the time-distance of one alternation of the ac cycle, the two alternations are identical except for polarity, and only one alternation occurs at a time. There are full-wave and multiple-wave antennas, but they are somewhat tricky to adjust. The half wave, in several variations, is more flexible for ham purposes. In actual practice, nearby objects affect the characteristics of an antenna and its performance and may require changes in its construction. They also alter the ability of the antenna to work slightly above and below its theoretical fundamental frequency.

The physical relationship between the half-wave antenna and the operating wavelength now becomes evident. For the popular 2-meter band, a half wave is only one meter long; that is, a few inches more than a yard. For 10 meters, it is 5 meters, or 16 feet 4 inches; for 20 meters, 32 feet 8 inches; for 40 meters, 65 feet 4 inches; for 80 meters, 130 feet 8 inches. The *meter* designations for the ham bands are approximate, and are rounded off for convenience. For example, the "40-meter" band, more specifically defined by the FCC as 7000 to 7300 kHz, runs from 41 to 42.8 meters.

The "meter" designations for the various ham bands, and their frequency range, are given in Table 7-1.

Table 7-1. Ham Bands

Band	*Frequency (MHz)*
80 Meters	3.5—4.0
40 Meters	7.0—7.3
20 Meters	14.0—14.35
15 Meters	21.0—21.45
10 Meters	28.0—29.7
6 Meters	50—54
2 Meters	144—148
1¼ Meters	220—225
¾ Meter	420—450

THE BEAM ANTENNA

An ideal half wave in free space tends to radiate more energy at right angles to its length than off its ends. This direction effect is readily increased by the addition of one or several wires parallel to the antenna and placed on one side or both sides of it. These extra wires are not usually connected to the antenna proper, which is known as the driven element, but they may be in certain instances. Depending on their length and spacing, the extra wires act either as reflectors of the signals from the driven element (in

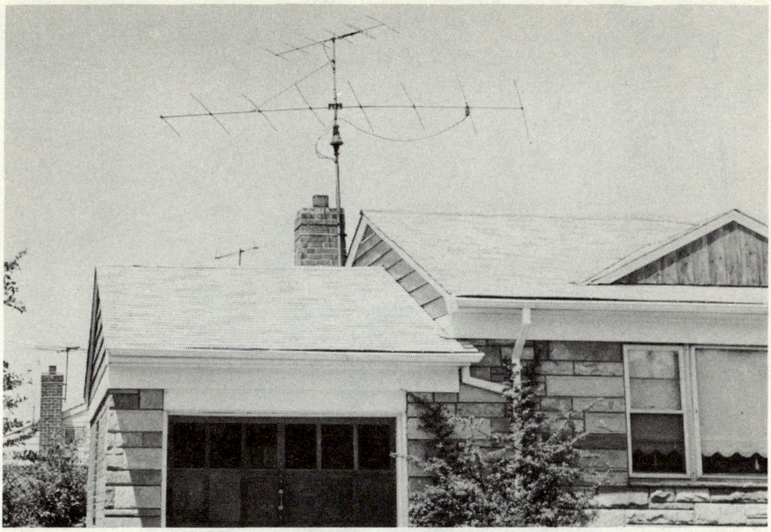

Here is a simple but highly effective combination of a 6-element beam for 2 meters (top of the pole) and an 8-element unit for 6 meters. In spite of its size and because of its light weight, this array is easily supported on the chimney and turned by an inexpensive tv antenna rotator.

the same manner as a polished bowl behind a lamp), or as directors, for the purpose of pushing as much of the energy as possible in a desired direction. Such an arrangement is called a *beam,* and is usually mounted on a tower or mast with an attached rotator to swing it around a full circle. In practical beams, the elements are not wires but lightweight aluminum tubing.

The greater the number of reflectors and directors aiding the driven element, the greater the concentration of signal energy along one path. This is equivalent to an enormous increase in *effective* radiated power for a given transmitter. The advantage is real, not merely theoretical.

The mere dimensions of basic elements for 40 and 80 meters make full-size beams for these bands somewhat impractical. For the shorter bands the elements are of reasonable size, and can be handled and erected without special equipment. Individual beams for 10, 15, and 20 meters usually have three elements. Some have four or five, a few have only two. Even a two-element beam is better than a naked half wave (also called a *dipole*), and it is cheaper and lighter than the multielement antennas.

Because it is expedient to shift from one band to another to take advantage of changing atmospheric conditions, by far the most popular of all beams is a three-element job that is made to work

Probably the most widely used tower-and-antenna combination to be found in the United States. The 15-foot tower supports the Mosley TA-33 beam. The objects near the ends of the horizontal elements are traps that permit the antenna to work on 10, 15, and 20 meters. The driven element is in the center; the reflector and director are on either side. The antenna mounted on the end of the vertical stick at the extreme left is a 2-meter ground plane antenna. It consists of a 19-inch center wire and four radials.

interchangeably on 10, 15, and 20 meters. The interchangeability is made possible by the addition of critically positioned coils called *traps* in each element. While the performance of individual antennas for each band is somewhat better than that obtained from the three-band antennas, the convenience afforded by the three-band model more than outweighs the extra performance.

For the 2-meter and shorter bands, the elements are so small and light that they can be piled on almost without limit. On 2 meters, it is common to hear hams using beams with six, eight, ten, twelve, and even more elements. On these high bands, still further *gain* (that is, increased effective radiated power over that possible with a dipole) is obtained by stacking several beams side by side and/or one over the other, and energizing them by the transmitter in such a manner that the elements radiate, reflect, and direct in unison. This is called *phasing* and becomes more difficult and critical as the number of elements and beams is increased.

On the fractional-meter wavelengths, which are the playground of experimentally inclined hams, radio waves begin to have some of the properties of light waves. A small radiator element can be placed at the focal point of a large reflector dish, usually made of open wire mesh or "hardware cloth." Even with very low power, such an electronic searchlight can beam signals over great distances.

FEEDING THE ANTENNA

Any radio antenna works best when it is located high in the air, out in the open. However, operating a radio shack from such a position would be most uncomfortable! It is necessary, therefore, to provide a transmission line of some sort to carry the signals from the transmitter to the antenna. There should be minimum loss of energy and little or no incidental radiation in this feeder to spoil the directional pattern of the antenna proper. These

A "quad" is an unusual antenna of wire strung on a flat rectangular frame. Here is a 4-element assembly made by Skylane Products, Temple Terrace, Florida 33617. Aerials of this type have high gain and are noted for their low angle of radiation, which means good DX transmission.

requirements are best met by coaxial cable; therefore virtually all modern ham gear is designed for it.

Coaxial cable or *coax,* as it is called, consists of an outer tubular shell of tinned copper wire braid, with a single inner conductor of either solid or stranded copper wire. The latter is insulated from the shell, in most cables, by a solid but flexible core of a plastic material. An extensive variety of plugs, jacks, and other fittings is available for making connections to the ends of the cable. The outer braid of most coax is covered by a plastic sheath to keep water from seeping in.

It is possible to use ordinary twin-lead television wire as the transmission line for simple dipoles. However, the characteristics of this material are quite different from those of coax. If the transmitter is intended for coax, a matching device called a *balun* must usually be connected between the transmitter and the shack end of the twin-lead.

The most efficient type of feeder is probably the *open line.* The open line consists of two bare wires kept uniformly parallel about an inch (2.5 centimeters) apart by means of slim plastic insulators which are spaced about every six inches (15.2 centimeters). It is available in rolls of 100, 250, and 500 feet (30.5, 76.2, and 152.4 meters). The disadvantages of open line are that it must be supported carefully with a minimum of bends, brought into the house through an insulated plate of some kind, and supplemented by a special antenna tuner.

HANGING THE SKY WIRE

Any antenna installation, especially the beam type, is much more of a physical than an electrical undertaking. It is easy enough to bolt a few aluminum tubes to an aluminum boom on the ground or on a roof, but the assembly becomes a real tiger to handle when you attempt to climb with it to the top of even a short supporting structure. You not only need three or four hands but also the ability to remain calm while the sky around you swings and sways—or at least you imagine that it does. It is normal for most people to become slightly dizzy at heights; therefore it is sheer suicide to go clambering over roofs and towers without adequate preparation and safety measures. Even a seemingly light 2-meter beam can become hard to manage in a breeze, and most larger beams can be put up safely only on dead-calm days.

If the roof of your house is readily accessible and you can stand comfortably near the chimney, make a modest start by using an 8- or 10-foot length of tv mast as support for a beam. With the assistance of one other person, you can readily walk the entire assembly up against the chimney and then secure it with regular tv-antenna chimney straps or brackets. The straps are preferable because they can be tightened quickly with a wrench and do not require preliminary drilling of holes in the brickwork.

If the roof is flat or only moderately pitched, an unguyed three-legged steel tower about 15 feet high is a safe project for three strong men. Plan its position very carefully so that it can be laid out flat with two of its legs loosely bolted to hinge plates, which,

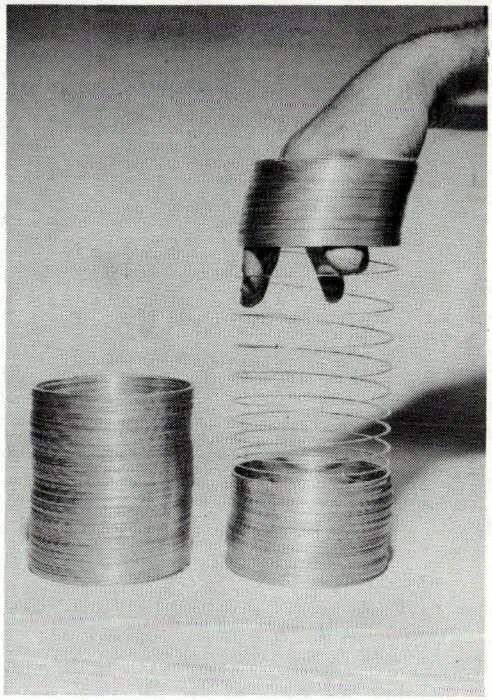

What is this "thing" doing in a chapter about antennas? It belongs because it is a clever antenna that adjusts to any band between 20 and 80 meters. "Slinky" is its name and it consists of a continuous helix of coated metal 4 inches in diameter and stretchable to 60 feet. It was developed originally for field use, but is also great for attics, small backyards, etc.

in turn, are secured to the roof itself. It is then easy to assemble the beam and its rotator to the top section. To keep the boom from hitting the roof during this operation prop the tower up with a box or a chair. With one man pushing, another pulling, and the third keeping the cables clear you can hinge the entire tower upward to its vertical position. Put in all remaining leg bolts, and the worst part of the undertaking is over.

Beyond about 15 feet, towers become too heavy and cumbersome for most roofs, and should be planned instead for ground mounting. The foldover type is by far the safest and most sensible one for a man who doesn't relish the idea of being a weekend steeplejack. With this design, the top folds down within easy reach of the ground; you can do any work on the beam in perfect safety. When you're finished, you merely crank the structure to the vertical position at any height you desire.

EXTENT OF DIRECTIONAL EFFECTS

The radiation from a beam is strongest in its directed position, but this does not preclude the possibility of appreciable radiation from its back or off its sides. You will frequently hear strong signals from stations whose beams, according to the operators' remarks, are pointed away from you. However, if you work one of these stations, the owner will surely swing his antenna toward you, and you will notice a startling increase in strength.

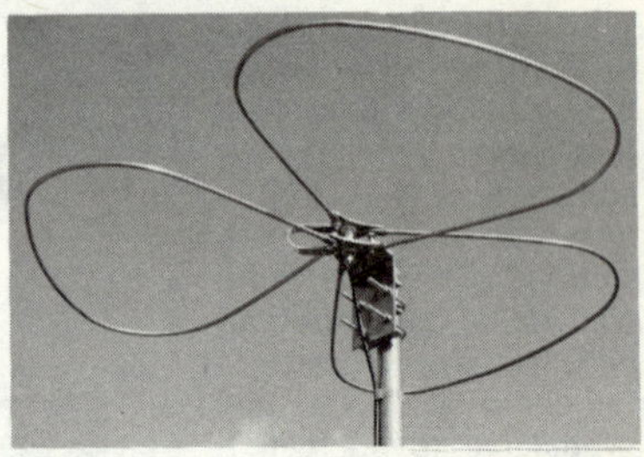

Here are two eye-catching antennas having omnidirectional characteristics; that is, they radiate equally well in all directions. These are the Cushman "Squalo" (left), available in five sizes for 6 through 40 meters, and the "Big Wheel" (right) in three sizes for ¾, 1¼, and 2 meters. Both types can be stacked for increased radiation.

Directional effects are much more marked on 2 and 6 meters than on 10 and upward. The reason for this is that the beams used for the shorter waves usually contain more elements than those for the longer waves, and the more elements, the more directivity.

On rare occasions your own beam, for receiving, may give the loudest signals from a particular station when it is pointed precisely 180 degrees away from what should be the correct setting. Before you climb up on the roof and start rebuilding the antenna, consider the possibility that the signals are arriving from the transmitting station the long way around the globe. This has been known to happen!

WIRE ANTENNAS

For 40 or 80 meters, a simple straight wire is usually quite satisfactory; about 65 to 66 feet for 40 meters, 130 to 133 feet for 80

This photo shows the "dream-of-a-lifetime" antenna of Nat Schnoll, W2PN, of Upper Saddle River, New Jersey. A three-section telescoping tower rises 71-feet high and is topped by a ten-foot mast. On the latter are two horizontal booms. The lower one is 43-feet long and holds a full-size 5-element beam for 20 meters. The upper one is 26-feet long and holds two separate 3-element beams, for 10 and 15 meters. Little wonder that W2PN is one of the world's leading DX operators!

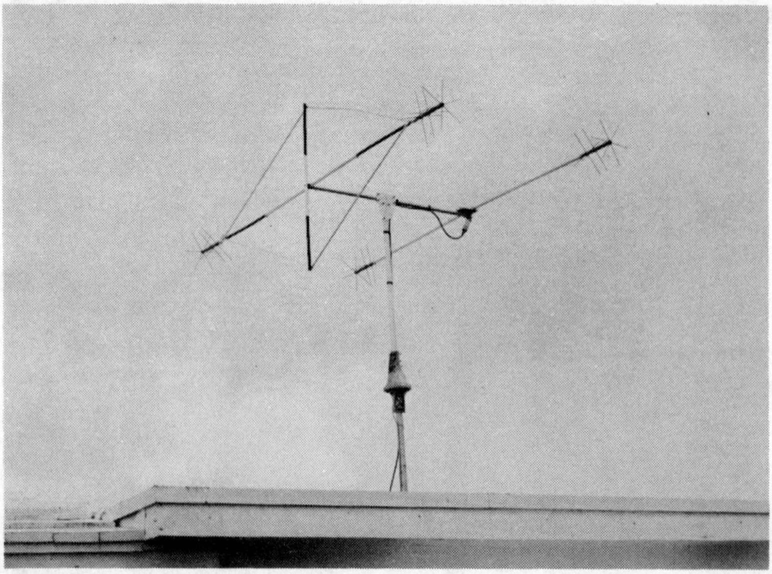

From the street this "Mini-Beam" appears to be growing out of the roof, but there isn't a single hole in the latter. The pipe is attached to a 2-foot square wood center platform, to which are bolted four legs made of 2 × 4s each 4-feet long and held down by concrete building blocks. The beam is 10 feet above the roof. The whole structure is an easy one-man job to assemble and to move for experimentation.

meters. These antennas can be trapped, in the same manner as beam elements, to work quite well on several bands. Long-wire antennas are very useful because they cost relatively little, are easy to suspend between house and tree or other supports, and are inconspicuous. The last consideration is important for apartment dwellers whose landlords don't take kindly to holes in the roof! A long wire antenna does not have the directivity and gain of a

With the addition of a "trap" in each of its halves, a simple wire antenna works well on 10, 15, and 20 meters. Coaxial cable at center goes to station. Supports are curtain poles on brick-weighted wood bases, held down by end guy wires to eaves of house.

beam, but if used intelligently, permits excellent long-distance communications.

VERTICAL RODS

A rigid aluminum tube about 30 feet (9.1 meters) long, mounted vertically on the side of a house, and fitted with an adjustable coil at its base, makes another simple and easily installed multiband antenna. It is even more convenient than a horizontal wire in crowded residential districts because it is confined to one's own property.

A vertical antenna is generally omnidirectional; that is, it radiates equally well in all directions. Its pattern may be disturbed slightly by nearby reflecting surfaces such as aluminum siding on a house or metal foil insulation in the walls, but the effect is comparatively minor. Like a dipole, the vertical has no built-in gain, but its own features make it very popular. Many hams use regular beams for the bands from 20 meters down and verticals for 40 and 80 meters. A coaxial switch in the shack makes the proper connections between the transmission lines and the transmitters.

8

Going Mobile

Not content to leave their radio equipment at home in the ham shack, many hams have equipped their cars, boats, and planes with amateur radio gear. What better way to travel than to have a friendly local ham guide you through a strange city? Also, what could be more enjoyable on a hot summer afternoon than sitting on a cool lake and talking to a fellow ham?

HAM SHACK ON WHEELS

Morton B. Kahn, formerly W2KR, Great Neck, N.Y., and now K4KR, Boca Raton, Fla., has long had a low-power transceiver in his car. One day, while inching his way through midtown New York traffic, he thought he'd check the functioning of a new tube that he put into the transmitter the night before. He flipped the switches on and as he watched the meters on the panel he spoke into his hand microphone just to activate the circuits.

> *"Hello CQ, hello CQ, CQ, CQ, this is W2KR mobile, William 2 King Roger mobile, on 20 meters, in New York City. Is this thing working, I wonder?"*

Half a second after he released the mike button the speaker on the dashboard came to life.

> *"Hello W2KR mobile, calling William 2 King Roger mobile, this is DL4HA, Dog Love 4 How Able, in Germany. You're putting in a swell signal over here, old man. How do you read me? W2KR mobile, this is DL4HA. Over.*

Thrilling? Enough to make your scalp tingle! Unusual? Not altogether. When conditions are favorable you can work all around the world with flea power and a buggy-whip antenna flapping on

Morton B. Kahn, K4KR, presses the button on his hand mike and his mobile transmitter is on the air. The unit suspended from the dash is his own design. Mort has worked many foreign countries on 20-meter phone.

your rear bumper. Conditions must have been good that day for Mort, because he repeated the performance, in reverse, when he retraced his route through Manhattan a few hours later. The ignition systems of passing trucks were creating a shower of interference in the receiver, but out of the uproar came a fairly good signal.

> *"Hello CQ, hello CQ, CQ, CQ, CQ. Calling stateside. This is CN8EY, CN8EY, Charlie Nan 8 Easy Yoke, in Morocco, Africa. What say some one please?"*

Morocco! Mort glanced at the walls of the canyon formed by the surrounding skyscrapers. "What can I lose?" he mused to himself as he reached for the mike.

> *"Hello CN8EY, CN8EY, calling Charlie Nan 8 Easy Yoke in Morocco, this is W2KR mobile; William 2 King Roger mobile, William 2 King Roger mobile, in New York City. How're we doing there, feller? Over."*

He learned soon enough how he was doing.

> *"Hello W2KR mobile, this is CN8EY. Thanks for the call, Mort. I've heard you on the air often and I'm glad to work you . . ."*

Fifteen minutes later Mort had to cut the contact because he had reached his destination. Neither station had missed a word.

As you might expect, mobile stations are especially active during holiday and vacation seasons. More and more, they are being installed in RVs (recreational vehicles). The first time you hear one, you might observe that its signal appears to be a bit unsteady and sometimes breaks off abruptly in the middle of a word. This is not necessarily the fault of the equipment; more likely it is due merely to the flapping of the antenna at high road speeds or to a self-imposed silence when the operator suddenly decides he'd better pay more attention to the traffic.

"Look ma, no hands!" Bil Harrison, W2AVA, believes in safe mobile operation, so he uses a neck-supported mike and a foot-operated switch.

The most popular mobile bands are 10, 15, and 20 meters. The 10-meter segment tends to be erratic, but when it does open up, because of changing atmospheric conditions, you can work all around the globe with very low power.

BOAT AND PLANE MOBILE

It is not generally known that mobile ham stations may be operated in boats and airplanes as well as in automobiles. The FCC regulations in this regard are rather simple:

(1) The installation and operation of the station must be approved by the master of the ship or the captain of the plane.
(2) The ham equipment must be separate from and independent of any radio equipment aboard the ship or aircraft.

(3) The ham equipment must satisfy safety regulations applicable to electrical apparatus, as promulgated by the appropriate government agency.
(4) The ham station must not interfere with other radio equipment aboard the vessel or plane.
(5) The ham equipment must not constitute a hazard to life or safety.

All FCC regulations regarding types of transmission, classes and privileges of operator licenses, power limitations, etc., apply to boat and plane operation as they do to fixed-station operation. You are free also to have on board a CB set and a marine radiotelephone, as they serve entirely different purposes.

This is a mobile ham shack. Where's the steering wheel of the car? There isn't any, because this isn't a car. Believe it or not, it's a submarine, the U.S.S. Cutlass, and its skipper, Commander J. Dan Reilly, USN, W4NMK, operates regularly UNDERWATER with a submerged antenna!

The value of the ham regulations is that they make amateur radio appealing to the hundreds of thousands of owners of personal pleasure boats. Since the owner of a small boat is also the "master," he can do anything he wants in regard to a ham station. The installations on some boats are very elaborate, even including beam antennas.

One big feature of mobile operation afloat is relative freedom from noise interference. It's an uncanny experience to listen on a high-gain receiver while you're anchored in a quiet cove and to note the absence of the usual racket caused by passing cars, leaky power lines, elevator switches, thermostat controls, loose electric bulbs, etc.

PHONE OR CW

Most mobile operation in cars is on phone, but you can also work CW with all its advantages when you are in a stationary position. Some experienced operators even work CW while in motion, although this sort of thing can be a traffic hazard. They use surplus army knee keys, which were designed for just this purpose, and they pound away merrily while they drive. One ham

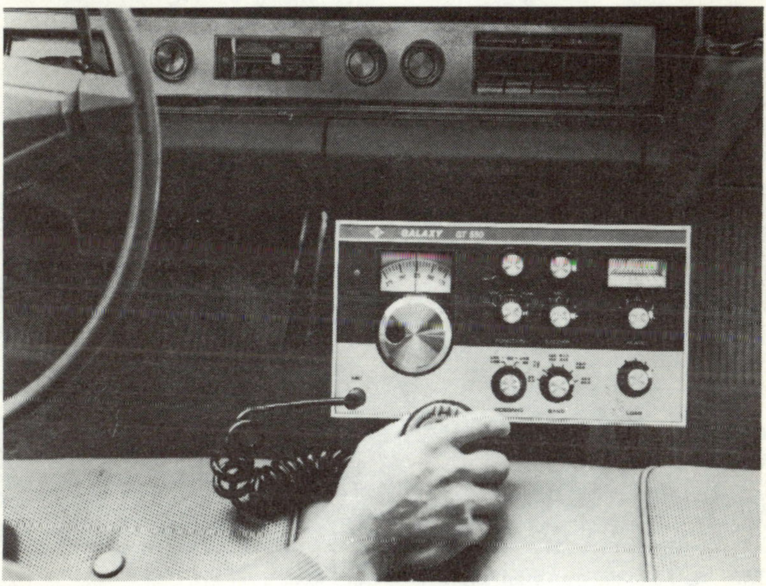

The Galaxy GT500 is a compact but powerful ssb transceiver covering the bands from 10 through 80 meters. Here it is installed under the dash for mobile operation; the 12-volt power supply is separate. With an ac power unit it works just as well in a fixed installation.

in Chicago operates mobile CW every morning on the way to work, and he has rolled up an impressive number of DX contacts with a transmitter putting out no more than *five* watts.

Mobile Equipment

The most popular mode of mobile operation is ssb, for the simple reason that it is also the favorite mode for fixed-station operation. It follows that the bands covered are the popular ones from 10 through 80 meters. All mobile equipment of recent vintage is of the transceiver type. Rapidly disappearing into home-station junkboxes are am converters that work through the am broadcast receiver in a car, am transceivers, separate am receivers and transmitters, and similar obsolescent equipment. (Two-meter fm transceivers are in a special category; see Chapter 12.)

All these transceivers are intended for mounting under the dashboard. To prevent them from interfering with the driver's accelerator and brake foot, they usually have to be centered over or moved slightly to the right of the hump for the transmission housing. In this position, they are likely to eliminate the leg room needed by a passenger occupying the center area of the front seat. Most drivers do not consider this a serious shortcoming, since three-abreast seating up front is generally a risky arrangement anyway. If the car has two bucket seats, the problem doesn't arise.

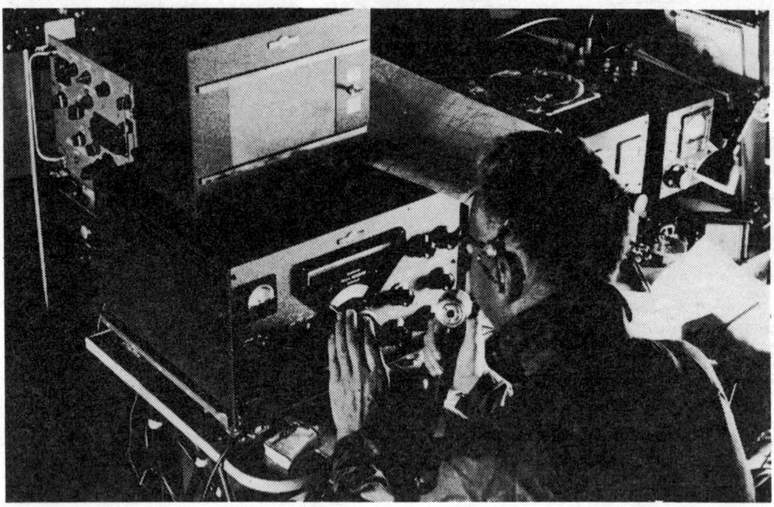

This is a ham station in an airplane. It does look a bit elaborate, but there is plenty of room for it because the craft is a huge Air Force cargo jet. Completely independent of the plane's regular communications and navigation equipment, the rig provides great DX opportunities for the many crew members who are amateur radio operators.

You can bet that her "CQ mobile" calls get plenty of answers! Mrs. Shirley Hargreaves, K2BPT, Clifton, N.J., enjoys using her Polytronics PC-6 transceiver because it looks pretty and puts out a strong signal.

Many mobile hams prefer a dash mounting bracket into which the transceiver can be snapped quickly and released just as quickly. With the unit hidden in the trunk of the car while it is unattended there is nothing to attract the attention of passing vandals. Most of them are looking for CB sets, but they usually don't know the difference between a CB and a ham transceiver.

All mobile equipment is designed for push-to-talk operation by means of a switch on the hand microphone.

Mobile radio is a special boon to residents of rental, cooperative, or condominium buildings on which outside antennas of any kind, even television, are strictly forbidden for aesthetic reasons. It usually does little good to argue that a line of black poles carrying long power lines all around the property is much uglier than three or four aluminum tubes on a thin pipe in the backyard.

Many hams overcome this problem by turning their cars into veritable stations-on-wheels. They are particularly happy if their wives or other companions on long trips do some of the driving, as this frees them to operate the equipment safely.

What makes mobile doubly interesting is that conditions change rapidly as you drive. In cities, one straight stretch of streets might be noisy and virtually empty of signals. Make a turn somewhere and the set comes to life with a bang. You might think that the best locations are the flat, unobstructed ones such as you find in the great plains states, along the shore lines of oceans, and on mountain passes everywhere. You can be fooled! You can park on a California cliff directly above the Pacific Ocean and shout yourself hoarse calling CQ without getting a single reply. On the other hand, you might sit in a motel parking lot deep in a mountain valley in Vermont and work station after station just after a single announcement that you are on the air.

It's worth remembering that mobile is open to all hams within the operating restrictions of each class of license.

"Hamming" afloat is a real joy, as well as a real comfort in times of emergency. And to do the job properly you must of course wear a white "skipper's" cap!

Aeronautical mobile is, literally, the highest form of amateur radio, as the author learned when he had the good fortune once to hitch a ride on a huge military cargo plane. It was fitted with a standard ham ssb receiver that did not interfere in any way with the other electronic gear on board. The transceiver could be used by crew members and authorized passengers who happened to be hams. The "chief operator" on this flight was a three-star general who had been a ham since boyhood. He turned on the rig while the plane was taxiing down the runway, and as the wheels were coming up he sent out a single CQ followed by his call sign. A lot of hams around the world must have known what his usual daytime frequency was because they came right back. Within ten minutes he had made rock-solid contacts with stations on both coasts, in several countries in Europe, in the Canal Zone, and with the ham captain of a freighter just west of Hawaii.

The regular commercial airlines don't even allow passengers to use small portable radio receivers on board, so of course it would be futile to ask permission to use a transceiver of any kind. However, at small airports where sightseeing rides are a regular weekend feature you will have little trouble finding a pilot who will take you up with a small portable transceiver that you want to try at three or four thousand feet. In fact, you might learn that he is a ham himself, and more than willing to cooperate. Pilots have to know quite a lot about radio to qualify for their pilots' licenses, so many of them drift into amateur radio almost without trying.

THE POWER PROBLEM

Until the advent of the transistor, mobile transceivers were essentially scaled-down versions of fixed-station equipment, used tubes, and required more power than many car electrical systems could furnish without going up in smoke. The tubes required not only heavy filament current but high plate voltage. Suitable power packs, using vibrators or dynamotors, were so bulky that they were usually stowed in the trunk of the car.

Transistors changed all of this for the better. Aside from the obvious advantage that they have no filaments to keep hot, they are essentially low-voltage devices that are easily satisfied with the 13.2 volts available from practically any car storage battery. In the receive mode, a typical solid-state transceiver draws about half an ampere; in the transmit mode, it draws five to seven amperes. This means that you can plug its power cord into the cigarette-lighter receptacle of the car without fear of overloading the

This Atlas Model 210 solid-state, four-band mobile transceiver is not much larger than some 5-watt CB rigs, yet it is rated at 200 watts input. In this typical installation, note how it mounts neatly near the steering wheel, with the microphone *on a clip next to it.*

battery. Whether you use a mobile transceiver or not, it is a smart idea to check the battery periodically with a hydrometer to be sure that it stays reasonably well charged with normal driving.

The dashboard ammeters in older cars are calibrated only approximately, and are hardly more than indicators of "charge" and "discharge." In most newer cars there is no ammeter at all, but only a red pilot light to show "discharge." Since it is important to know what goes on in the electrical system of the car, you should install a good zero-center ammeter reading up to 50 or 100 amperes. Mount it on the underhanging lip of the dashboard.

Auxiliary Power Source

A picnic in the country becomes doubly enjoyable if you park your car in some quiet, secluded spot, preferably on a hill, where reception and transmission are both good. However, if you let the engine idle fast, to power the transmitter and conserve the battery, it is likely to overheat.

Hams who go in for mobile in a big way often invest in a one-cylinder gasoline-engine-driven 12-volt generator. Small enough to be carried in the trunk and to be handled by one man, this generator runs most of a day on a couple of gallons of fuel. Setting it up and clipping its leads into the car takes about five minutes.

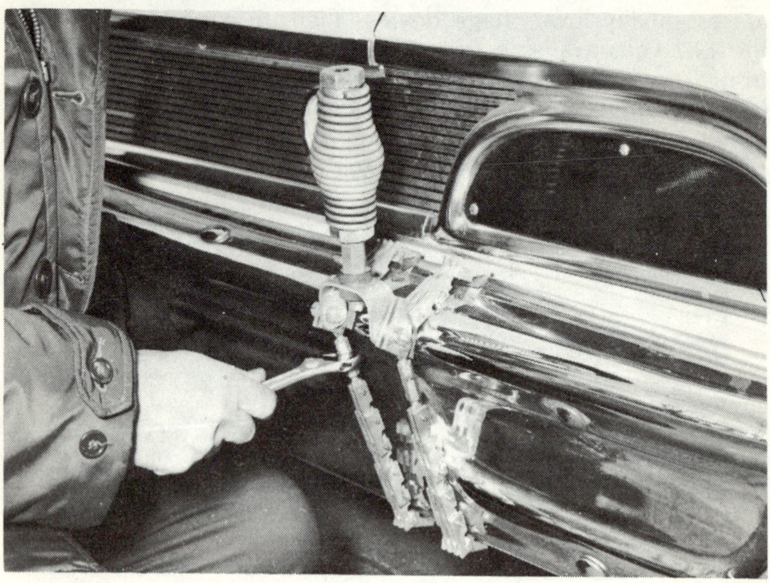

Bumper mount eliminates the need for cutting holes in the body.

You'll often see one chugging away at ham field days, model airplane meets featuring radio control, sporting events where hams provide the point-to-point communications, etc. Generators of this type are widely used in rural areas, camps, etc., as a primary source of power.

For use at the events mentioned above, hams often carry their regular home-based, ac-operated equipment, set it up under tents or other shelter, and power it from generators that put out 115 volts ac. This is known as "portable" rather than "mobile" operation. The arrangement is very popular because it frees the car for its normal transportation purposes. A typical portable generator with an output of 350 watts weighs only about 14 pounds and is the size of a small suitcase. A larger unit is rated at 1250 watts, weighs 65 pounds, and can be carried by one strong man or two weaklings.

Some generators produce 12 volts dc at about 10 amperes in addition to the ac. This can be very useful for running battery-operated model transceivers or for charging car batteries.

Most small- and medium-size ac generators have rope-pull starters, as on outboard engines for boats. Some larger models can be started electrically by means of jumper cables to a car. Once the generator unit is running, the jumpers are disconnected and the car is free again.

THE INTERFERENCE PROBLEM

In older cars, interference created by the electrical system can make reception on the ham bands very difficult, although it usually doesn't show up at all on the broadcast band. The ignition elements usually get most of the blame, but actually ignition interference is rather easy to cure by means of a suppressor resistor in the central distributor lead and bypass capacitors on the regulator and generator terminals. The worst trouble sometimes is caused by such seemingly innocent devices as the fuel gauge or clock. You may have to spend quite some time trying small mica capacitors between ground and the terminals of all fixtures in the electrical system. The queerest type of interference comes from incomplete grounding between various large members of the car itself, especially if the engine is mounted on large rubber shock blocks.

At high road speeds, in dry weather, the friction of the tires against the road bed generates old-fashioned static that sounds like a thunderstorm. This interference is noticeable even on the broadcast band. Hams have been known to borrow an idea from large fuel trucks; they attach a short length of chain to the bumper

The Mosley MA-3 antenna works on 10, 15, and 20 meters without switches or adjustment—a great convenience in mobile operation. The owner of this car didn't mind making a hole in the body to put the antenna in the clear. Result: He works worldwide DX with a 50-watt transceiver!

and let it trail over the ground. The static charges that otherwise would run all over the car are thus grounded off, literally.

Late-model cars are gratifyingly "clean" of internal interference because the bonding operation between chassis members is carried out in manufacturing and is quite thorough. It is not unusual now to find the engine block connected to the fire wall by a length of one-inch wide copper braid. There is often a wiping contact arm where the hood closes against the body.

SAFETY

Mobile radio is fun, but so is living. When traffic conditions are bad, leave the rig off and concentrate on driving. Save mobile operation for the open road where there are fewer distractions. Another person in the car is, of course, a great help in keeping the log and making minor adjustments on the transmitter. Ideally, let

This bumper-mounted coaxial-type 2-meter antenna belongs, obviously, to K8BLL, Alan Robertson, Benton Harbor, Michigan. Onlooker is K8ADS.

With a bumper sticker like this one over a distinctive license plate, you are sure to be asked: "What's a ham?"

someone else drive so that you can devote full attention to your operating. Remember also that all the rules that apply to fixed stations apply to mobile as well.

MOBILE EQUIPMENT

Ham shacks on wheels are becoming so popular that some manufacturers are concentrating on mobile gear. Representative examples of mobile equipment have been pictured throughout this chapter. As for other types of equipment, consult the free manufacturers' or distributors' literature for complete descriptions, prices, and specifications for a specific piece of equipment.

MOBILE ANTENNAS

A simple 8-foot whip antenna is a quarter-wave long at 10 meters; in combination with the metal body of the car, it acts as an efficient radiator. For the 15-meter and higher bands, a straight whip becomes impractical, because it hits against trees, overhead wires, etc. The usual expedient then is to insert loading coils in a short whip to make it work as a longer one. These coils can be found at the base or at the center. If two loading coils are carefully located at critical separations, a single antenna can be made to work without adjustment on three bands, usually 10, 15, and 20 meters.

No wonder K6OHJ is happy! He drives a Corvette fitted with a mobile rig and a Webster Band-Spanner antenna. The cutaway view shows the internal inductance adjustment for precise setting to the desired frequency.

What's the picture of a living-room ham shack doing in the mobile chapter? Answer: This is not a living room. It's the salon of the 83-foot ocean-going yacht "Compromise" and the National equipment is the maritime mobile station of Peter Schweitzer, W2DMQ.

In the Webster *Band-Spanner* center-loaded whip the loading coil is wound directly on the fiber-glass support column. A whip of fixed length is arranged to slide in and out of the open neck of the column. Its lower end carries a circular contactor that touches against the exposed turns of the coil on the inside. Raising or lowering the whip, plunger fashion, enables the user to adjust the antenna very accurately. The range of variable inductance introduced by the loading coil covers 10 through 80 meters.

A completely different approach is represented by the Mark Mobile *Heliwhip*. The designers felt that ordinary whips are too long, too unwieldy, and too conspicuous, so they produced a much shorter one, made of fiber glass. This insulating material is only the support for the actual radiator element, a spirally wound wire of variable pitch. A *Heliwhip* for 6 meters is only three feet long; for 10 to 15 meters, four feet; and 20, 40, or 80, six feet. There is also a three-band (10, 15, and 20 meters) *Heliwhip* six feet long.

Mounting a mobile antenna is a simple mechanical operation. If you don't mind making a hole in the body, the best place is the flat part of the rear deck or a fender. Excellent results are also obtained from whips mounted on the rear bumper.

9

How to Be a Good Operator

After you pass your license test and before you put a transmitter on the air, you must take a little time out to acquaint yourself with ham operating procedure and practices. This is necessary whether you start with CW or phone, and doubly important for CW because you have to express yourself by tapping a key instead of merely talking into a microphone. Many a ham who rushes on the air with a new license and new equipment gets himself into an awful stew during his first contact when he finds that he just can't understand what the other fellow is saying. This is very frustrating and can take all the fun out of the contact.

The formal rules are prescribed by the FCC. In addition, common use is made of many informal abbreviations and expressions which are special to the ham game and are utterly incomprehensible to outsiders.

THE MEANING OF *CQ*

By far the most commonly used and heard expression on the ham bands is *CQ*. This combination of letters is the ham's way of announcing to anybody who happens to hear him that he is on the air and is anxious to make a contact. The origin of the signal is somewhat vague. It is probably a contraction of *CQD,* the distress call used in the early days of radio before the adoption of the international distress signal *SOS*. CQD was popularly interpreted to mean "Come Quick Danger," and of course operators hearing it would give it their concentrated attention.

CW OPERATION

In CW work certain special signals are required to make the intention of a transmission specific and unmistakable. One that is used in every call is the intermediate *DE,* which is taken from French and means "from." It is placed between the call letters of the called station and the call letters of the calling station. Military stations usually use the letter V for the same purpose.

Good operating is made easier if you can read all of the dials on your equipment without either eye strain or neck strain. A series of blocks under the front edges of receivers and transmitters bring the panels up to an easy viewing angle. Open space on table is handy for log sheets, Call Book.

There are five ending signals, the applications of which are illustrated later in this section. The single letter *K* terminates a CQ call and a transmission to a station with which definite contact has been made. It says, in effect, "Go ahead. I'm finished for the moment, and I expect an answer."

The signal *di-dah-di-dah-dit,* which is represented in print as $\overline{AR}$[1], is used at the end of a call to a specific station before con-

[1] The overscore indicates that the dots and dashes are sent with uniform spacing, and are not separated into two letters.

tact has been established. For instance, if you heard a station send CQ, you could call it, follow with DE and your own call, and $\overline{\text{AR}}$ as the termination. The $\overline{\text{AR}}$ is also considered a sort of invitation to other operators to call you in case the station you called does not come back to you. $\overline{\text{AR}}$ is used at the end of a formal message, as distinguished from the usual "rag chewing" in which hams engage.

$\overline{\text{KN}}$ is a variation of "go ahead." It is used at the end of a transmission when an answer is wanted only from the station called, and no other. *K* by itself really has the same significance, but under some conditions it isn't strong enough to discourage other stations from joining a contact. $\overline{\text{KN}}$ is particularly useful after directional CQ's, during emergency communications, etc. The *N* added to the *K* is like making a warning sign read "POSITIVELY No Smoking."

$\overline{\text{SK}}$ is the same as "good bye." It means that you are finished with the station you have been working and that you are ready for any other business that might present itself. If you are finished completely for the evening, sign off in addition with *CL*. This is not a manufactured signal like $\overline{\text{AR}}$, $\overline{\text{KN}}$ and $\overline{\text{SK}}$, but is sent as two separate letters and is short for "closing."

Don't bother to write down ending signals. Just follow their meaning.

Special Signals and Punctuation

The letter *R* used by itself is the signal of receipt. It means, in effect, "I have received and I understand all of your last transmission." Don't send *R* and then ask for repeat of transmission.

A string of *V*'s, is the standard test signal. Use it when you are tuning up the transmitter. Other operators hearing the *V*'s will know what you're doing and will not waste time trying to call you.

$\overline{\text{AS}}$ is another useful signal. It means "wait," and is not to be acknowledged by the receiving operator. You might be hunting a fresh pencil or lighting a cigarette; send a few $\overline{\text{AS}}$'s to let the other chap know you're still with him.

$\overline{\text{BT}}$ is a dash, a separating signal between the call letters of a transmission and the words or messages you want to send. It is something of a stalling signal too. If you haven't quite decided what to send after finishing one thought, tap off a couple of $\overline{\text{BT}}$'s. They represent a shorter interval than $\overline{\text{AS}}$ and help to keep the contact alive. The $\overline{\text{BT}}$ is frequently used instead of the period between words of the usual rag chew.

As mentioned in Chapter 2, the period and the comma are rarely heard in informal interchanges among hams. However, they exist and you are at complete liberty to use them any time you think they will clarify the meaning of something you want to send. The question mark, on the other hand, is in very common use, particularly in conjunction with the Q signals. The question mark also has the meaning "repeat," in the sense of a request when an operator misses part of a transmission and wants it sent again.

Abbreviations

The passion for abbreviation in CW operating has been carried by some hams to the point of absurdity. Instead of simplifying the exchange of intelligence, which is the basic intention, the practice confuses beginners and old-timers alike and leads to many requests for repeats. In most cases, less total time would be consumed on the air if the whole words were sent in the first place.

For your guidance in decryptographing some of the gibberish you're likely to catch on the air, there follows a list of frequently heard abbreviations, with comments attached to some of them. If an unknown one is thrown at you, don't hesitate to ask the other operator to switch to English!

Abbreviation	*Meaning*
AA	All after.
AB	All before. (These two are used with the question mark to get a repeat of sections of a message. They are followed by the nearest words of the text, as identification.)
ABT	About.
ADR	Address.
AGN	Again.
BCI	Broadcast interference.
BCL	Broadcast listener. (Always used in a slightly contemptuous sense to distinguish a mere receiver owner from a ham.)
BK	Break. (This signal is usually used to interrupt or break the other operator's sending, so that you can tell him something important.)
BN	Been. (Or "between," used like AA and AB for getting repeats of message text.)
C	Yes. (This is baffling. If the simple three-letter word is to be abbreviated at all, why not be logical and make it Y? The C is probably a carryover from old telegraph practice, when it was used to mean "correct" after an operator queried the spelling of a word.)
CK	Check. (Refers to the word-count of a formal message.)
CL	Closing down station.
CLD	Called.
CLG	Calling.
CUL	See you later. (This is a compound abbreviation, the C and "see" being the same phonetically.)
CU AGN	See you again.
DX	Distance. (An old ham abbreviation that got into the popular language during the radio boom of the late 1920s and early 1930s.)
FB	Fine business. (A common expression of approval or praise.)
FREQ	Frequency.
GA	Good afternoon. (Also sometimes improperly used instead of K.)

Abbreviation	*Meaning*
GB	Good bye.
GE	Good evening.
GG	Going.
GM	Good morning.
GN	Good night.
GND	Ground.
HI	This is not a greeting, like "Hi, there!" but is best described as a Morse laugh. It's a short way of expressing amusement.
HR	Here or hear.
HV	Have.
HW	How.
MSG	Message.
N	No.
NCS	Net control station.
ND	Nothing doing.
NIL	Nothing, or I have nothing for you.
NR	Number.
NW	Now.
OB	Old Boy.
OM	Old man. (Regardless of his age, every male ham is automatically an OM. It's a friendly, fraternal form of address.)
OP or OPR	Operator.
PSE	Please.
RPT	Repeat.
SKED	Schedule.
TFC	Traffic.
TMW	Tomorrow.
TNX-TKS	Thanks.
TT	That.
TU	Thank you.
TVI	Television interference.
UR-URS	Your, Yours.
VY	Very.
WA	Word after.
WB	Word before.
WRD	Word.
WL	Well, will.
WX	Weather.
XMTR	Transmitter.
XTAL	Crystal.
XYL	Ex-young lady. (Wife!)
YL	Young lady. (Male hams don't mind being called "old man," but imagine any female ham, or any female for that matter, *not* objecting to "old lady" or "old woman"!)
73	Best regards. (This is an old telegraphic sign-off.)

THE INTERNATIONAL "Q" CODE

The first great use of radio was of course on ships at sea. Its rapid and widespread adoption gave rise to a serious language problem. How, for instance, could the Spanish operator of a freighter bound for Seattle send and receive vital navigational information to and from the American operator on shore? To meet this situation, one of the early international radio conventions adopted a series of three-letter signals, all starting with Q and having the same meanings in all languages. While designed primarily for maritime use, many of them are sufficiently general to be valuable also for ham purposes. With their aid, you can exchange enough intelligence with another ham in Brazil, for example, to make the contact an interesting one.

Each Q signal has both an affirmative or answer meaning and also an interrogatory meaning, the latter being formed merely by the addition of the question mark after it. You'll find a complete list in the Call Book. Reproduced here with comments are the signals having definite amateur applications. The comments in parentheses are the author's.

Abbreviation	*Question*	*Answer*
QRG	Will you tell me my exact frequency?	Your exact frequency is . . . . . . kHz.
QRH	Does my frequency vary?	Your frequency varies. (Before telling this to a ham, make sure it's not your receiver that's drifting.)
QRJ	Do you receive me badly?	I cannot receive you. Your signals are too weak.
QRK	What is the legibility of my signals?	The legibility of your signal is . . . (In ham practice, the answer to QRK? is usually expressed according to the R-S-T scale, described in this chapter.)
QRL	Are you busy?	I am busy (or I am busy with . . .) Do not interfere.
QRM	Are you being interfered with?	I am being interfered with (QRM has gotten to be a common abbreviation for "interference.")
QRN	Are you troubled by atmospherics?	I am troubled by atmospherics. (That is, "static.")
QRO	Shall I increase power?	Increase power.
QRP	Shall I decrease power?	Decrease power.
QRQ	Shall I send faster?	Send faster.
QRS	Shall I send more slowly?	Send more slowly.
QRT	Shall I stop sending?	Stop sending.
QRU	Have you anything for me?	I have nothing for you.
QRV	Are you ready?	I am ready.
QRW	Shall I tell . . . that you are calling him on . . . kHz?	Please tell . . . that I am calling him on . . . kHz.
QRX	Shall I wait? When will you call me again?	Wait (or wait until I have finished communicating with . . . I will call you at . . . o'clock, or immediately).

Abbreviation	*Question*	*Answer*
QRZ	Who is calling me?	You are being called by . . . (QRZ? is very useful after you call CQ and can't quite get the call of an answering station.)
QSA	What is the strength of my signals?	The strength of your signals is . . . (Comment on QRK also applies here.)
QSD	Is my keying correct; are my signals distinct?	Your keying is incorrect; your signals are bad. (If more hams used this signal, instead of being polite liars, CW operation would improve greatly.)
QSL	Can you give me acknowledgement of receipt?	I give you acknowledgement of receipt. (This somewhat duplicates the simple R for "received." See last paragraph, this chapter.)
QSM	Shall I repeat the last telegram I sent you?	Repeat the last telegram you have sent me. (Why send "pse rpt ur last msg," as some hams do, when QSM says all this in three letters?)
QSO	Can you communicate with . . . direct (or through the medium of . . . ?)	I can communicate with . . . direct (or through the medium of . . .) (Hams invariably refer to a successful contact with another station as "a QSO.")
QSV	Shall I send a series of V's?	Send a series of V's.
QSY	Shall I change to transmission on . . . kHz without changing the type of wave?	Change to transmission on . . . kHz without changing the type of wave. (Changing frequency is thus often called "QSYing.")
QSZ	Shall I send each word or group twice?	Send each word or group twice.
QRRR	This is a special signal, and if you hear it keep off the frequency except to listen, unless you are in a position to help. It is the official ARRL land distress call, for emergency use only. It is the equivalent of SOS as used by ships at sea, and must receive the same attention and priority.	

SIGNAL REPORTING

Every ham wants to know how well he is being received, so one of the first things he asks, after making a QSO, is QRK? The answer to this question is an estimate expressed in the *R-S-T* system. R stands for readability, and ranges from 1 for "unreadable" to 5 for "perfectly readable." S stands for signal strength, and ranges from 1 for "faint signals, barely perceptible," to 9 for "extremely strong signals." T is for tone, and 1 means "extremely rough hissing note," 9 for "purest dc note." Since there is no standard method or yardstick for measuring or judging any of these characteristics, any signal report is only a friendly approximation on the part of the receiving operator. Two people listening to the same signal on the same receiver often will not agree on an R-S-T report.

Hams generally tend to inflate the value of a signal, to make the other guy feel good. Sometimes they make themselves sound slightly foolish, as when they say, "Ur sigs FB hr OM RST 489 QRM bad QSZ." Translated, this means, "Your signals are fine here, old man. Readable with practically no difficulty, strong, purest dc note, but interference is bad so send each word twice." Don't give this type of report!

THE FIRST QSO

By this time perhaps you're itching to warm up your transmitter and go on the air. Let's assume that you have a novice license and will work initially on CW, which is a good way to get experience. You have pencils, a pad of paper, a Call Book, etc. You have a choice now: Send your own CQ and fish around for an answer, or listen for a CQ from another ham. Be a sport and start the ball rolling yourself. A CQ must be fairly long, to give other hams the chance to tune into it. By "long" is meant a full minute. If this doesn't sound like much, try timing it against the sweep second hand of a clock. The best system is call-three-sign-three, like this:

CQ CQ CQ DE WN1ABC WN1ABC WN1ABC CQ CQ CQ DE WN1ABC WN1ABC WN1ABC

(Keep repeating, and at the end of the minute finish with K.)

Luck is with you, and you hear your own call letters coming back at you:

WN1ABC WN1ABC WN1ABC DE WN4XYZ WN4XYZ WN4XYZ wot say OM? $\overline{AR}$

You thumb through the Call Book hurriedly. "Wow, he's in Florida!" Now what are you going to say to him? Here's a suggestion:

WN4XYZ WN4XYZ DE WN1ABC WN1ABC R $\overline{BT}$ *Tks for the call OM This is my first QSO and am sure happy* $\overline{BT}$ *ur RST 479 hr in Manchester N H* $\overline{BT}$ *Hw WX down there? WN4XYZ DE WN1ABC K*

Note that the call up at the beginning can now be reduced to two-and-two, since your signals are already tuned in. Also, the identification of both called and calling stations is repeated once at the very end of the transmission. The terminating signal is now *K* because contact has been made. The answer might be:

WN1ABC DE WN4XYZ R $\overline{BT}$ WI OM FB Glad to be ur first QSO WI send QSL card $\overline{BT}$ WX hr plenty warm went swimming this morning Hws skiing up there HI $\overline{BT}$ nice signal hr in Miami RST 469 QRM as usual WN1ABC DE WN4XYZ K

If conditions are good you'll gossip back and forth for maybe ten minutes, and then:

WN4XYZ DE WN1ABC R $\overline{BT}$ WI OM my XYL says chow is ready so must QRT Hope to work you agn and tnx for invite to see shack 73 $\overline{SK}$ WN4XYZ DE WN1ABC CL

The reply will be short:

WN1ABC DE WN4XYZ R Good appetite 73 to XYL too $\overline{SK}$ WN1ABC DE WN4XYZ

Now you can pull the switches, but before you leave the radio table don't forget to enter the required data in your log book. The time of starting and finishing, call of other station, frequency and power used, and signal reports are all entered. You'll really enjoy dinner now!

As you listen more and more, you'll realize that the "end of work" signal $\overline{SK}$ is somewhat abused and misused. $\overline{SK}$ means that you're finished, through, terminated, wound up; you not only don't expect but don't want further conversation other than an acknowledgement. If the other chap ignores the $\overline{SK}$, you can be polite and go along with him, but if you have to do something else you may have to just turn off the set. If the other guy decides to end the QSO, take the hint from his 73 and $\overline{SK}$ and do likewise.

PHONE OPERATION

The basic procedure used in CW work applies to phone too, but of course is simpler because you can talk naturally. Only a few standard expressions are needed, as follows:

The equivalent of CQ is "Calling any ten-meter phone, calling any ten-meter phone." (Use appropriate frequency or wavelength figure.) A variation is, "Hello CQ, CQ, CQ, calling any ten-meter phone, calling any ten-meter phone."

The intermediate signal is merely "this is" or "from." Following a CQ, "this is" leads into your own call letters a little better. "From" is usually used in a short-form call after contact has been made, as illustrated later. In any event, the sequence of call letters must be preserved: first the call of the called station, then "this is" or "from," and then the call of the calling station. The reverse form, "WN1ABC calling WN2DEF," is contrary to FCC regulations.

The ending signals are "go ahead" or "over," and "signing off" or "out." The first two mean, "My transmission is ended and I

An open U-shaped arrangement of equipment on a desk top puts everything within comfortable reach of the operator. He is Morton B. Kahn, K4KR, Boca Raton, Florida. His assistant, Caesar, is a real radio hound.

expect a response from you." The second two mean, "This conversation is ended and no response is expected." "Over" and "out" are part of a compromise United States-British military procedure developed during World War II and still largely used.

The universal voice signal of receipt or acknowledgement, understood even where American isn't, is "Okay." You will hear a great many hams using the word "Roger" in the same sense as Okay. This is a funny case. "Okay" had been used previously in U.S. military voice procedure, but the British objected to it as not "proper." As part of the compromise, then, the phonetic equivalent of the telegraphic signal of receipt, the letter R, was chosen; this was Roger in the World War II phonetic alphabet, and it stuck.

Phonetic Alphabets

In voice operation, it is often difficult to distinguish between *B*, *C*, and *V*, between *D*, *P*, and *T*, and other letters. The FCC says in its regulations, "When using telephony, phonetic aids to identify the call sign of the station may be employed." Unfortunately, the FCC does not prescribe an official phonetic alphabet, which amateur radio sorely needs. Hams are free to use anything they like, and the result on the air is often confusion rather than clarification.

Probably the most generally used phonetic alphabet is the one that grew out of the aforementioned United States-British military agreement. Hundreds of thousands of men and women in the services were trained to use it for ordinary telephone communication as well as radio, so the carryover to ham operation was more or less natural. More recently, a new international alphabet was adopted by the military services and commercial airlines. The two are given side by side in Table 9-1, with the words of the new

Table 9-1. Phonetic Alphabets

		New International	
Letter	*Old Military*	*Word*	*Pronunciation*
A	Able	Alfa	AL-fah
B	Baker	Bravo	BRAH-voh
C	Charlie	Charlie	CHAR-lee
D	Dog	Delta	DELL-tah
E	Easy	Echo	ECK-oh
F	Fox	Foxtrot	FOKS-trot
G	George	Golf	GOLF
H	How	Hotel	HOH-tel
I	Item	India	IN-dee-ah
J	Jig	Juliet	JEW-lee-ett
K	King	Kilo	KEY-loh
L	Love	Lima	LEE-mah
M	Mike	Mike	MIKE
N	Nan	November	No-VEM-ber
O	Oboe	Oscar	OSS-cah
P	Peter	Papa	Pah-PAH
Q	Queen	Quebec	Keh-BECK
R	Roger	Romeo	ROW-me-oh
S	Sugar	Sierra	See-AIR-rah
T	Tare	Tango	TANG-go
U	Uncle	Uniform	YOU-nee-form
V	Victor	Victor	VIK-tah
W	William	Whiskey	WISS-key
X	X-ray	X-ray	ECKS-ray
Y	Yoke	Yankee	YANG-key
Z	Zebra	Zulu	ZOO-loo

international alphabet spelled out phonetically for clarification. The stress is on the capitalized parts.

The numbers are spoken in the usual fashion, except that 9 is "niner" and the cipher is "zero," never "oh." To distinguish zero from the letter O (Oscar), particularly as used with the call signs of stations in the tenth district, it is usually written as Ø.

Voice Distress Signal

The international voice distress signal is the word "Mayday." It's the phonetic equivalent of the last word of the French expression *Prière de m'aider,* which means "Please help me." In every sense, Mayday means the same as SOS. It's used mainly by aircraft, and also by small vessels equipped with ship-to-shore radiophones.

On the Air With Phone

Phone operation is just talking, so it should be easy. Wrap yourself around the mike and see what you can do. Again start with a general call:

> *"Hello CQ, CQ, CQ, calling any 20-meter phone, local, or DX. CQ, CQ, CQ, calling any 20-meter phone, calling any 20-meter phone, calling any 20-meter phone. This is W2DUX, William 2 Dog Uncle X-ray, in Hamville, Long Island. CQ, CQ, CQ, calling any 20-meter phone. This is W2DUX, William 2 Dog Uncle X-ray, William 2 Dog Uncle X-ray. What say someone please. Over."*

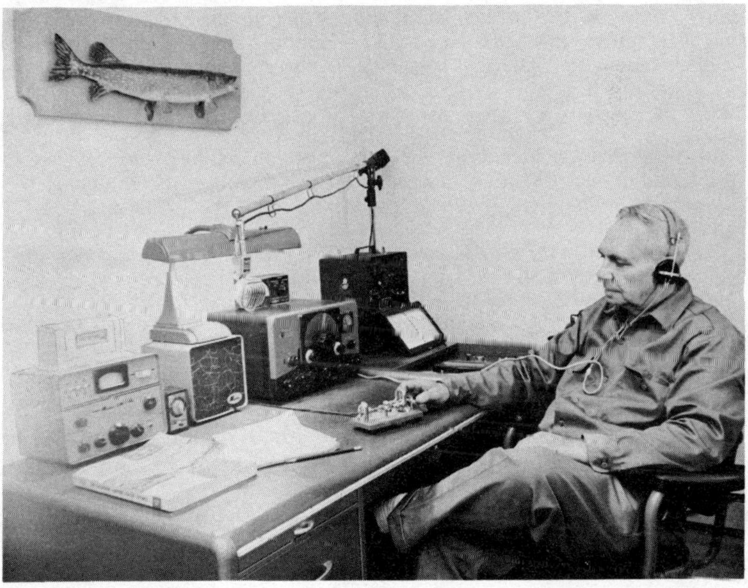

A standard office desk and a matching swivel chair provide solid comfort for Carl Mosley, WØFQY, St. Louis, Missouri. Equipment (left to right): Mosley CM-1 receiver, speaker, Viking Ranger transmitter, BC-221 frequency meter, Viking one-kilowatt linear amplifier. Carl rattles a mean "bug" and uses phone too. Northern Pike, on wall, represents his other hobby.

Do not say "Charlie Queen" for CQ, as CQ by itself is the most often heard and easily recognized expression on the air. Talk slowly and distinctly for about a minute, and tune around for an answer. You might hear:

"Hello W2DUX, W2DUX. Calling William 2 Dog Uncle X-ray, William 2 Dog Uncle X-ray. This is W2GLP, William 2 George Love Peter, W2GLP, William 2 George Love Peter, in Sunnyside, Long Island. Do you read me, old man? W2DUX, this is W2GLP. Go ahead."

Now you pick it up:

"W2GLP, W2GLP, William 2 George Love Peter. This is W2DUX. Thanks a lot for the shout. Just got this rig working and you're my first QSO. Handle here is Paul. You're putting in a very strong signal. Anxious to know what this thing sounds like. Will you please give me a check? W2GLP from W2DUX. Over."

Since contact has definitely been established, the call-up is shorter than initially, and elaborate use of the phonetic alphabet is no longer necessary because the stations have identified themselves correctly.

"W2DUX from W2GLP. All okay, Paul. Handle here is Al, Able Love. You won't have any trouble getting out with that transmitter. Good loud signal, but I think you're overmodulating a bit. On your next transmission you might try turning down the audio gain a little, and I'll be glad to give you another report, Paul. W2DUX from W2GLP. Over."

This can go on for an hour. Finally you decide it's time to quit.

"W2GLP from W2DUX. Well, Al, you've been a great help and it's been a pleasure working you. Have to shut down and take the wife shopping, so if you have nothing further I'll be pulling switches. W2GLP from W2DUX. Over."

"W2DUX from W2GLP. Glad to be of help, Paul. You run along. Will see you some more. W2DUX, this is W2GLP, William 2 George Love Peter, signing off and clear."

"So long, Al. W2GLP, this is W2DUX signing off and pulling switches."

Note the difference in sign-offs. By identifying himself with phonetic words and saying ". . . and clear," W2GLP is indicating to others who might be listening that he is remaining on the air and would welcome other calls. The last phrase ". . . and pulling switches" means that you're positively going off the air. Also note that the call-up is omitted from the last transmission, because of its brevity.

Break-in Procedure

There are times when you'll hear two other stations with great strength and clarity and you'll be sorely tempted to try to cut in and make it a three-way conversation. This can be done if it's done carefully. Wait patiently for a moment when both stations are silent, then turn on your transmitter quickly and snap out, "Break break."

If one of the other operators hears you and is willing to be interrupted, he'll say, "Who's the breaker?"

With this invitation, you proceed. "W6PXH, this is K2DUX. Thanks for letting me in. Want to let you guys know you're pounding in like a ton of bricks here. . . ." Once you've been accepted, you carry on normally.

If the other operator hears you but isn't quite ready for you, he'll say, "Stand by the breaker." This is your cue to keep quiet and wait until he returns eventually with, "Now who's the breaker who wanted in?"

There is the possibility that your interruption isn't welcome, in which case the other chaps will discourage you by not answering at all. The one kind of communication you should never interrupt is phone patching.

Traffic Handling

Since by definition an amateur station is one operated as a hobby, it follows that hams are prohibited by law from transmitting messages for other people for pay. Within the United States and to its possessions and territories, and to scattered military establishments, they can handle "third party" traffic just for the fun of it, and many do. Most such messages are of the innocuous "love and greetings by amateur radio" type, and if they take a week to go 500 miles no one suffers any inconvenience.

Third-party traffic between the United States and many foreign countries is prohibited altogether because it is considered unfair competition for the commercial communications services. Look in any of the monthly ham magazines for current lists of "open" and "closed" countries.

The Phone Patch

One of the truly great public services performed by ham radio is the almost daily hooking-up of servicemen overseas and their families at home, usually by way of the 20-meter phone band. This is accomplished at the domestic end with the aid of a *phone patch* —a simple coupling device between the ham rig and the telephone. While phone companies are usually fussy about their instruments and strongly dislike extra connections on the lines, they more or less overlook phone patches, for a good reason: they are responsible for many calls that otherwise would not be made at all.

A phone patch puts telephone hookups on the air, a great stunt for letting service personnel overseas talk to their families stateside. This Heathkit unit Model HD-15 connects to the line without upsetting the characteristics of the ham equipment.

Let's take a common situation. An American soldier in Germany visits a ham station run by GIs—probably a MARS unit—and asks if there's any chance of getting a call through to Louisville, Kentucky. "We can try," is the usual answer. The operator goes on the air:

> *"Hello CQ Louisville, Kentucky or vicinity only, for a phone patch. CQ Louisville, Kentucky or vicinity only, for a phone patch. This is DL4ABC, Dog Love 4 Able Baker Charlie, in Germany. CQ Louisville, Kentucky or vicinity only, for a phone patch. This is DL4ABC, Dog Love 4 Able Baker Charlie, in Germany. Over."*

These directional CQs always get attention because hams know how important they are to homesick lads in strange lands. If conditions are good the answer might be:

> *"DL4ABC, DL4ABC, Dog Love 4 Able Baker Charlie. This is W4DEF, William 4 Dog Easy Fox, William 4 Dog Easy Fox. I read you loud and clear here in Louisville and will be glad to work a phone patch. DL4ABC from W4DEF. Over."*
>
> *"W4DEF, W4DEF. This is DL4ABC. Thanks, old man. Here's the dope: Corporal Johnny Jones calling his mother, Mrs. Mary Jones, phone number BLuegrass 9-3568. Got it? W4DEF from DL4ABC. Over."*

"DL4ABC from W4DEF. Okay. Stand by while I call."

The ham in Louisville quickly calls the number. "Mrs. Jones? This is Bill Williams, on Derby Road. I'm an amateur radio operator, and I'm talking to a station in Frankfort, Germany, where your son Johnny is standing by to talk to you."

The usual reaction to this sort of thing is consternation, delight, or hysteria, or a mixture of all three. A wise parent or wife will compose herself and be cheerful.

"It's extremely kind of you to go to this trouble, Mr. Williams."
"Not at all, Mrs. Jones. I have to switch back and forth, as only one of you can talk at a time. He'll talk first. I may have to cut in to keep you from getting mixed up, but don't mind. Now please wait and take your cue from either Johnny or me."
Shifting to his ham mike:

"Hello DL4ABC, DL4ABC. This is W4DEF. All set at this end. Johnny, take it away!"

"Hello mom, how are you and pop? I'm feeling fine . . ." Magic words worth more than anything.

The first time you tune accidentally into a phone patch you'll feel slightly embarrassed, but you'll also feel proud that ham radio can do such a marvelous job of it. One of the prize hook-ups of recent times was between a newly arrived airman in North Africa and his young wife in a tiny town deep in the Southwest. She had given birth to a son shortly after her husband had shipped out. As she spoke softly into the phone, obviously fighting to control herself, the baby could be heard very distinctly, gurgling and making other little sounds. This went on for about fifteen minutes before the band pooped out. It is doubtful if there was a dry eye in any of the numerous ham shacks around the world where the conversation was picked up.

KEEPING TIME

When association between even adjacent communities was limited, time was a matter of purely local concern. Towns, cities, and whole countries set their clocks any way they wanted, with the high-noon sun as the approximate standard. However, with the development of improved means of travel and communication during the second half of the 19th century, the need for an international time base of some kind became urgent. The principal commercial nations of the world finally agreed on a system called *Greenwich Mean Time,* or *GMT* for short.

The earth was scored off, like an orange about to be peeled, into 24 segments each 15 degrees wide. The starting point for time calculations was then established as the imaginary line that runs from the North Pole to the South Pole through a marker at an astronomical observatory in Greenwich, England. This "zero" meridian is the center of the Greenwich zone, which, like the other segments, is 15 degrees wide.

Greenwich was probably chosen because time is important in navigation at sea and Great Britain was the leading sea power of the period. The observatory was almost destroyed by German bombing during World War II, but the actual marker has survived.

For each 15-degree zone east of Greenwich (which is part of London) the local time is one hour ahead of Greenwich time; for each zone west, the local time is one hour behind. Thus, 75 degrees or five zones west of Greenwich, across the Atlantic, are such cities as Boston, New York, Washington, and Miami. Closer to Greenwich to the east are Paris, one hour ahead, and Athens, two hours ahead.

These two pictures show how the 24-hour time system eliminates confusion. The round 12-hour clock in both scenes reads 9:24, but you can't tell just by looking at it whether this represents morning or evening. There's no doubt with the digital readout clock on the left, a Numechron-Tymeter. The reading 0924 can mean only morning, and 2124 can mean only evening anywhere.

In recent years there has been a movement away from the designation *Greenwich Mean Time* and toward *Coordinated Universal Time*. The latter is now used exclusively by the Bureau of Standards for WWV and its other time-signal stations, which can be heard all over the world. The expression is often abbreviated to *UT* or, in some services, merely to the letter Z. However, GMT has

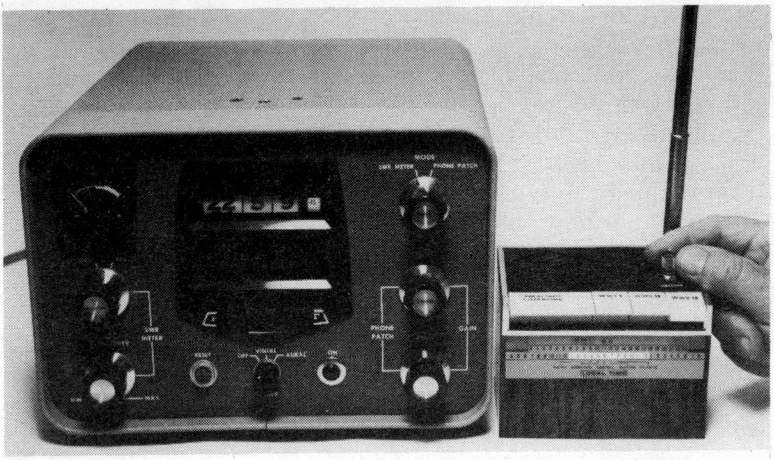

If your receiver or transceiver doesn't have a position for station WWV, you are missing the most accurate time signals in the world, transmitted daily by the Bureau of Standards. What you need is a Radio Shack "Timekube." Powered by a single transistor-radio battery, this clever little device offers three push-button frequencies: 5, 10, and 15 MHz. Push a button and the signals pop out.

Hard-earned but unusual wallpaper! The late Dr. Harold H. Riker, K2JHA, Flushing, New York, had a prize collection of QSL cards as a result of his hamming on ssb. The cards on this wall are all foreign; the ones covered by his left hand are from France, Italy, Germany, and Sweden.

been used for a century and probably will not fade out for years to come.

In the GMT concept, the abbreviations AM and PM and the term *o'clock* do not appear at all. All GMT times are expressed in four digits, the first two representing the hour and the second two the minutes past the hour. The conventional morning hours from midnight through noon are thus 0100, 0200, etc., to 1200. The first hour past noon is 1300, and the sixth hour plus 34 minutes past noon in 1834. It is impossible to confuse 2:00 AM with 2:00 PM if they are expressed in the 24-hour system, because the first is 0200 and the second is 1400 hours. When using these figures verbally, follow the system used by the announcers at WWV. For times before noon: *seven hours, thirty five minutes; nine hours, four minutes,* etc. For times after noon: *thirteen hours, six minutes; twenty hours, twelve minutes,* etc. When writing the times always use four digits and add GMT, UT, or Z: Ø735, Ø9Ø4, 13Ø6, 2Ø12. (Ø, not 0, is zero.) Avoid "Eight hundred hours" and the like; remember, the day has only 24 hours.

QSL CARDS

After working each other for the first time, most hams exchange QSL cards. These are usually regulation postcard size, 3¼ by 5½

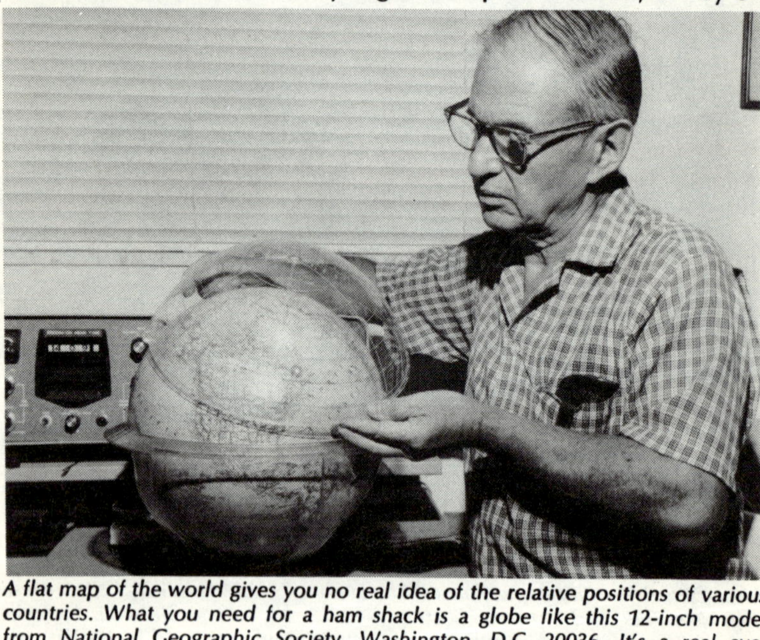

A flat map of the world gives you no real idea of the relative positions of various countries. What you need for a ham shack is a globe like this 12-inch model from National Geographic Society, Washington, D.C. 20036. It's a real eye-opener!

inches. On one side is a description of the station, with the call letters printed in large, bold type. Numerous small printers who make a specialty of these cards advertise in the ham magazines. Many cards are very elaborate and make interesting wall displays, and the stamps alone on some of the foreign ones are worth having.

To prevent cards from going astray, make sure of the other fellow's name and address by looking in the Call Book while you are still in contact with him, and have him verify the listing. If you can't find a listing, ask him for all particulars and give him your pedigree in return.

Remember, when sending a card to a foreign ham, *don't* put a 10-cent stamp on the card and expect it to reach its destination. Check the postal regulations for the proper postage.

10

Test Equipment and Safety Measures

A multirange meter that measures dc and ac voltage and resistance is the most useful piece of test equipment for ham purposes (for all electronic test purposes, for that matter). Such meters are available in two forms. The first is known as a *multimeter* or volt-ohmmilliammeter, abbreviated vom. It consists of a low-current range dc ammeter (a microammeter or milliammeter), an assortment of small resistors, a rectifier, a couple of flashlight cells, and a selector switch with as many as eighteen positions. The switch throws the resistors into various series or parallel combinations with the meter to make it read in millionths of an ampere (microamperes), thousandths of an ampere (milliamperes), or whole amperes; dc volts, or ac volts with the rectifier cut in; and resistance with the batteries connected in. A vom is self-contained and requires no outside source of power. Connection between it and the circuit or device under test is made by a pair of long flexible wires, the free ends are fitted with heavy insulating handles as protection against the voltage encountered in some equipment.

The second type of measuring instrument bears the designation *electronic voltmeter.* In its modern form, it is entirely of solid-state design (usually incorporating field-effect transistors, or "FETs"); it is more versatile than the basic vom; and it operates on self-contained batteries independently of the ac line. In its older form, it uses vacuum tubes and is therefore called a *vacuum-tube voltmeter,* or vtvm; and it works off the ac line. Both styles are available in both factory-made and kit forms, at a wide range of prices. The kits are easy assembly jobs.

METER SENSITIVITY

Mutimeters are described as having a sensitivity of so many *ohms-per-volt*. Small, inexpensive meters are usually 1000 ohms-per-volt; the better grades, used for ham and tv-radio service work, are rated at 20,000 ohms-per-volt. What this means is that the effective resistance represented by a meter when it is connected across a circuit is equal to the scale range multiplied by the sensitivity rating. For example, if the selector switch of a 20,000 ohms-per-volt meter is set to 10 VOLTS DC, the instrument acts as a fixed resistor of 200,000 ohms in addition to functioning as a 10-volt voltmeter. This relatively low value of resistance upsets the normal operation of some circuits and may cause the meter to give false or misleading readings. On the higher ranges the effect is less marked because the resistances are higher. On the 300-volt scale, for example, the value is 300 times 20,000, or 6 megohms.

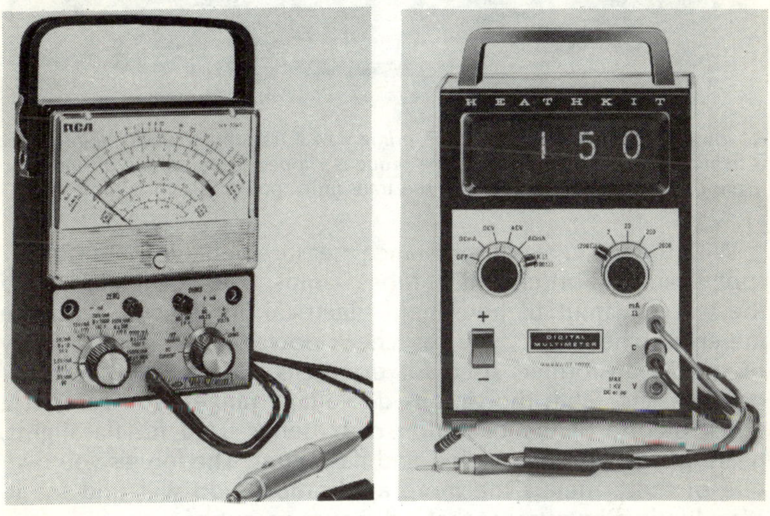

Representative electronic voltmeters popular with hams and technicians. Left: RCA Model WV-500B "VoltOhmyst" solid-state meter works on self-contained batteries. Right: Heathkit Model IM-1202 is an interesting multimeter with digital readout instead of a scale.

The big advantage of the electronic voltmeter is that it has a high input resistance, usually 11 megohms, on all voltage ranges. It is extremely valuable for measurements in delicate circuits where the voltages are low. Its resistance ranges are usually better spaced out on the scale and are more flexible than those of vom's.

Which type to choose? Many hams solve the dilemma in a simple manner: they buy one of each! It is often important to read voltage in one part of a circuit and current in another; use the electronic voltmeter for the first job and the vom for the second. It is often convenient or absolutely necessary to have separate meters to check two voltages at the same time, to see the effects of certain adjustments. For outside work, as on a mobile rig in a car, either a vom or an FET meter is satisfactory.

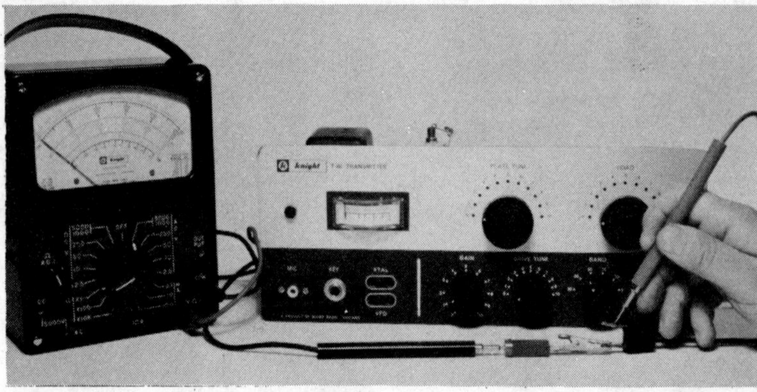

A typical vom application. Here the primary of a transmitter power transformer is being checked for continuity. One probe is clipped to one prong of line plug, the other to second prong, and the transmitter power switch is turned on.

With the meter set for resistance measurement, you can quickly spot open or short circuits in tubes, lamps, connecting cords, and the whole gamut of household electrical appliances. With the meter switched to ac, you can check power wiring and a variety of small transformers used for doorbells, toy trains, thermostat controls, etc. With the various dc voltage ranges, you can get a quick picture of the condition of batteries used for flashlights, hearing aids, portable radios, and flash guns. The longer you own one of these meters the more applications you will find for it. It's a lifetime investment that will never go out of date.

TRANSMITTER METERS

Virtually all transmitters, even the simplest ones sold in kit form, are provided with a milliammeter. It will have two or three scales and is switched into various parts of the circuit to show the currents in them. Some of the more elaborate transmitters have several milliammeters, which give simultaneous readings to facilitate tuning operations. Virtually no transmitters include volt-

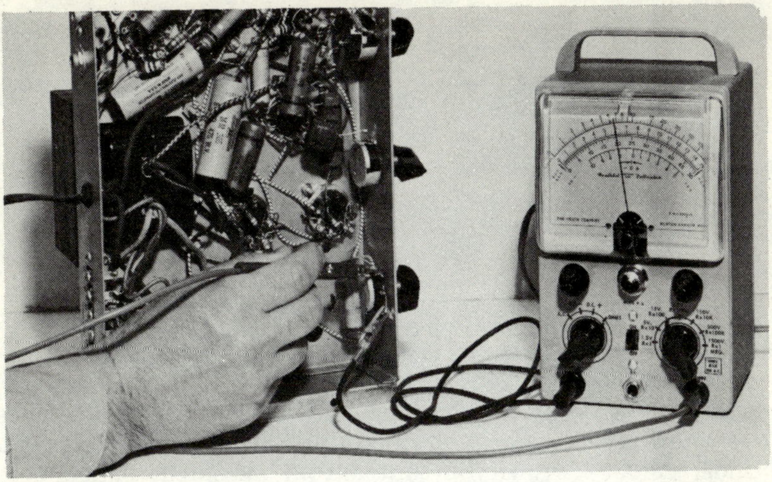

A vtvm being used to check the resistance value of a volume control, and also its condition. Irregular meter readings, as control knob is rotated, means that the resistance element is worn; this is a probable cause of noise.

meters. If you want to measure the power input to the final amplifier, connect the negative lead of your vom or FET meter to the chassis of the transmitter and the positive lead to either terminal of the milliammeter. Set the latter to read final-amplifier plate current and fire up the rig. Multiply the reading in volts by the reading in amperes and you have watts. Remember that a milliampere is 1/1000 ampere, so move the decimal point three places to the left when you make a notation of the plate current. For instance, 135 mA is .135 ampere. If the plate voltage is say 430, the power input is 58.05 watts.

DANGER—HIGH VOLTAGE

Make no mistake about it—even the smallest ham transmitter is dangerous to human life. It's dangerous in the same sense that a bathtub, a kitchen stove, or an automobile is dangerous. If you have any brains at all, you climb in and out of a slippery tub slowly and carefully. You open the gas cocks only if the pilot light is working or if you have a lighted match in hand. You apply the brakes gently when you're driving on an icy road. In other words, you're careful.

The fundamental rule of all electrical safety is: *Don't touch any exposed metal object on a receiver or transmitter with your bare fingers until you are sure it's "cold."* How can you be sure it's cold and not "hot"? Not by looking, but by testing. Don't

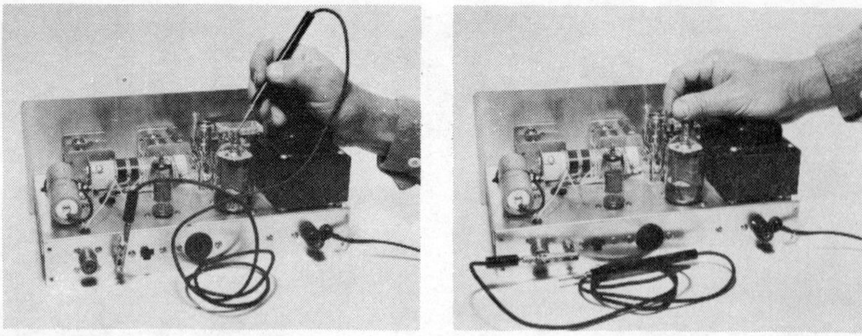

Wait! Before touching plate cap of transmitter tube, make sure that it isn't hot. Clip one end of a single test lead to the chassis and touch probe to cap as shown at the left. The plate cap can now be removed in complete safety.

even trust the on-off switch of the equipment. Most such switches are single-pole controls and open only one side of the ac line. Under certain peculiar but not uncommon circumstances, the other side of the line may still be hot in relation to ground and to you too if you happen to insert yourself in the circuit. The next rule is: *Pull out the line plug*. But wait a second . . . you're not in the clear yet. The large filter capacitors used in the power

Bleeder resistor cannot always be trusted. Play safe and short-circuit all filter capacitors to chassis with a single test lead before touching them.

supply sections of both receivers and transmitters can hold a heavy charge of electrons. In well-engineered equipment there are "bleeder" resistors in the circuit that function to dissipate any charges remaining in the capacitors after the power has been turned off. Because these resistors invariably are of high value, they cause the electrons to dribble off slowly, much as a small diameter pipe restricts the flow of water from a large storage tank. It can take the capacitors three to five seconds to discharge. In electrical practice that is a long time. If you turn the line switch

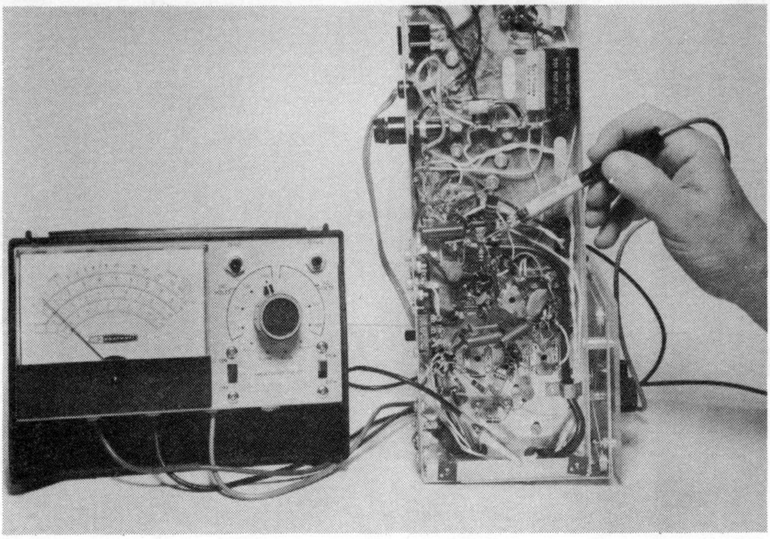

The proper one-hand method of making voltage checks in a live piece of equipment. The negative test lead is clipped to chassis and the positive probe is held near its end. This is a sure way to avoid being shocked by dc voltage. Instrument is Heathkit IM-17 solid-state (FET) voltmeter.

off with one hand while you reach to make an adjustment with the other, you may be in for quite a surprise, a shocking surprise, in fact! You can see for yourself how slowly some capacitors discharge after primary power is removed. Merely connect your voltmeter between B+ and ground. Turn the equipment on for a minute, then turn it off. The meter needle will not snap back to zero, but will drift down lazily. The third rule of safety then is: *Discharge possible charged capacitors.* The easiest way to do this is with a test lead from your vom or FET meter. Connect one end, by means of a clip, to the chassis, and tap the insulated probe at the other end to any or all terminals, soldering lugs, bare wires, and tube caps, that you expect to touch. The full discharge of the filter capacitors used in transmitters can be as loud and as startling as that of a .22 caliber cartridge. With the equipment disconnected from the power line and the capacitors emptied out, you can now stick your fingers anywhere. The only other caution you should observe relates to hot tubes, especially rectifiers.

Bleeder resistors are known to burn out, and therefore bleeders are not to be trusted 100% as safeguards against capacitor shocks. In fact, because they are across the highest voltage in a set, they take a continuous beating and are more susceptible to trouble than many smaller resistors in other circuit positions.

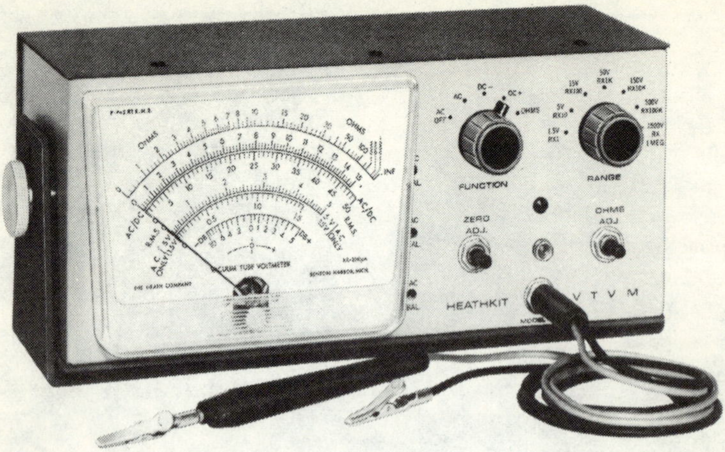

Horizontal-style vacuum-tube voltmeter, exemplified by Heathkit Model IM-28, is handy for bench use. Viewing angle is adjustable.

High-grade capacitors can hold their charges astonishingly well for long periods. The author has good reason to know this! He turned off a transmitter one day, with everything apparently in working order. Three days later, he reached in to change the final amplifier plug-in coil. The next thing he knew was that he was in a heap on the floor eight feet from the radio table, with a badly stung right hand and a lump on the back of his head! Sure enough, a large wirewound resistor, installed just to prevent such an accident, proved to be completely open.

Another good hint when working around high-voltage circuits is to keep one hand in your pocket. This way, you cannot accidentally touch the chassis while your other hand may be in contact with a voltage.

WATCH THE POWER LINES

Hams who show the proper respect for high-voltage dc sometimes make the mistake of underestimating the primary 115-volt ac line. If your hands have perspired or are damp or wet, and many times when they're bone dry, this much voltage can do serious damage and is often fatal. *Never* attempt to change or add wiring in outlet boxes, lamp sockets, extension cords, etc., without first removing the fuse in the cutout box. This may cause a bit of temporary inconvenience by darkening the room, but it's better to work by flashlight for a few minutes than to be encased in permanent darkness six feet under.

BODY RESISTANCE

Most people think of copper wire and other metals in general as good conductors of electricity. They overlook the fact that under the right circumstances the human body is a good conductor. You can make some interesting and significant investigations along this line with your vom or FET meter set on one of its higher resistance ranges. Hold the end of one test lead between the thumb and forefinger of your left hand. Touch the other test lead to various parts of the left hand, the elbow, the shoulder, the neck, etc. Grasp the metal tip of this lead tightly and loosely with your right fingers. Try wetting the skin with plain water, soapy water, water with a pinch of salt in it. The meter needle will dance around crazily, and the readings may be as low as several hundred ohms, and as high as several megohms.

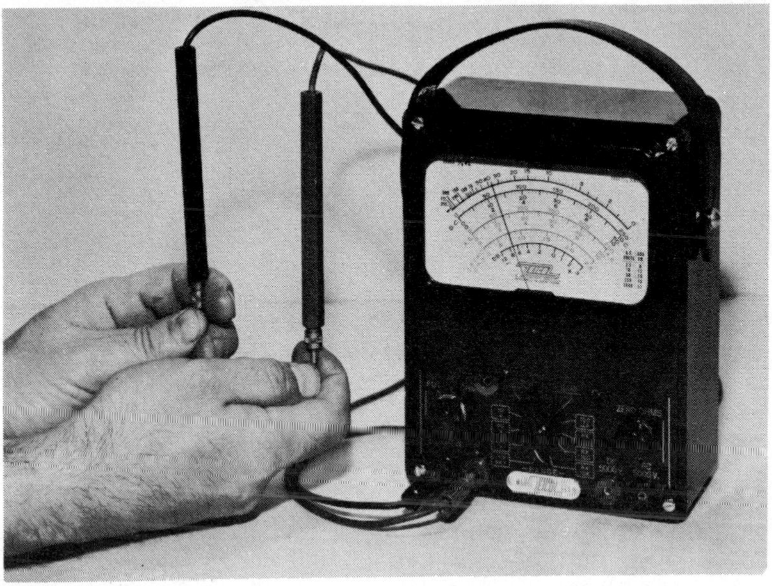

How good a conductor are you? Here the resistance between the two hands is only about 350,000 ohms.

WHAT VOLTAGES ARE DANGEROUS

You can put your fingers directly on the terminals of a 22½-volt battery, and feel either nothing or only a very faint tingle. With a 45-volt battery you'll get a slight twinge. Most normal people start to get stronger reactions beginning with 50 or 60 volts. Just keep clear of anything higher.

11

The Organization of Amateur Radio

There are hundreds of ham radio clubs, scattered all over the country. Many of them have permanent "homes" containing meeting rooms, code practice tables, workshops, and elaborate club stations. New members are always welcome and are quickly made to feel that they "belong." By all means join a club if there's one in your vicinity. You can determine if there is a club in your vicinity by looking for announcements of meetings in your local newspaper or on the bulletin boards of radio jobbers, etc. If you write directly to ARRL headquarters (see page 162), they'll tell you where the nearest one is located.

Strongly organized clubs can do a great deal to protect and promote ham interests on the local level. For example, the adoption of ham call letters for automobile license plates in many states is the direct result of club effort.

As an indication of the durability of ham radio as a hobby, it is interesting to note that the largest single club in existence, with a membership of about 14,000 in all call areas and in several foreign countries, is the Quarter Century Wireless Association. The only hams who are eligible are those who have held licenses continuously for a minimum of twenty-five years. There are regular meetings of chapters in many parts of the United States. For further information, write to the Q.C.W.A., Inc., 1409 Cooper Drive, Irving, Texas 75061.

THE *ARRL*

Since 1914, the amateur radio enthusiasts of the United States have been banded in a single national organization called the American Radio Relay League, or ARRL. The word "relay" is part of the name because in the early days of the game, before the DX possibilities of the short waves were understood, messages between distant stations had to be retransmitted by a number of intermediate relay stations. Practically all active hams are members or at least follow news of ham activities and developments in the league's publication, *QST*. At one time this was an official "Q" signal and meant "General call to all stations," so it is quite appropriate as a magazine name. The annual membership dues are $12.00 and include a year's subscription to *QST*. An interest in amateur radio is the only requirement for membership. Ownership of a station is not necessary, but the privilege of voting in

Sports events and parade traffic are handled with ease in Cleveland when Walter Ermer's ham radio corps assists police by providing auxiliary communications. Ermer, W8AEU (standing), is shown here during a sports car race discussing a traffic problem with George Flemming, K8QPH.

elections is extended only to holders of valid amateur licenses. The headquarters of the ARRL are at 225 Main Street, Newington, Connecticut 06111.

The league is divided into 15 geographical divisions, each with an elected director and a vice-director. The officers of the ARRL itself are also elected individuals. No one commercially engaged in the manufacture, sale, or rental of radio apparatus is eligible for the positions, so the league over the years has been able to live up to its stated purpose as a "noncommercial association of radio amateurs, bonded for the promotion of interest in amateur radio communication and experimentation, for the relaying of messages by radio, for the advancement of the radio art and of public welfare, for the representation of the radio amateur in legislative matters, and for the maintenance of fraternalism and a high standard of conduct." The league's representatives are familiar and welcome figures at FCC hearings and various other national meetings in Washington. They have also participated in many important international conferences pertaining to worldwide amateur activities.

PUBLIC SERVICE

Hams spend a great deal of their time on the air chewing the rag and adjusting their equipment. But when disaster strikes in

During the Cuyahoga river flood sixteen families were rescued by Ted Posey (right), one of Cleveland's 300-man amateur radio emergency corps.

the form of floods, storms, hurricanes, earthquakes, forest fires, etc., they perform outstanding public service by providing what is often the only means of communication in the stricken areas. Their extraordinary record for helping to save lives and property is recognized in official circles and greatly enhances their standing. In fact, the FCC has issued special regulations formalizing this activity, which is popularly called RACES, for *Radio Amateur Civil Emergency Service*. Although the overall Civil Defense program of the country has been dragging its feet, because of public

The Cleveland police often use the ham emergency corps directed by W8AEU (right). Here police Lieutenant Machael Roth, himself a ham (he is K8KNJ), uses one of Ermer's homemade portable transceivers.

apathy, the radio facilities furnished by the ham fraternity are quite efficient and prove their value every time a public emergency arises. This is acknowledged in one of the annual reports of the FCC, which states:

"Public service performed by 'hams' during weather emergencies such as snowstorms and hurricanes and also in times of fires, accidents, etc., is a continuing record. A large part of this service is through the Radio Amateur Civil Emergency Service (RACES) for the benefit of community civil defense organizations. RACES functions during peace to aid in times of natural disasters and emergencies. In wartime when other amateur activity must cease, RACES is the means whereby amateur activity may continue to furnish essential communication to help civil defense authorities."

One of the features of ham conventions is a flea market, which naturally attracts bugs—that is, radio bugs like the one who has his call on his shirt. This is no joke, he really is W4BUG!

CONVENTIONS AND HAMFESTS

Numerous conventions and hamfests are held by large clubs, state organizations, and ARRL divisions. These affairs usually include excellent technical talks, demonstrations of new equipment, displays by manufacturers or jobbers, code contests, fashion shows for the ladies, luncheons and dinners, etc. A national convention is held every few years. All are enjoyable, and give hams a chance to meet other hams they have worked on the air. The standard means of identification is a QSL card or an engraved name plate pinned to the coat or shirt.

12

2-Meter FM

Short-distance voice communication on the 2-meter (144–148 MHz) band with elementary tube-type am transceivers was long popular with amateurs as a change from DX hunting on the 10-, 15- and 20-meter bands. However, the shift from am to fm, the appearance of far superior solid-state transceivers, and the adoption of entirely different operating methods have revolutionized this aspect of the ham hobby. In fact, 2-meter fm is virtually a whole new game, even for many members of the Quarter Century Wireless Association.

The advanced techniques developed for the 144-MHz band have been extended also to the 220- and 440-MHz bands.

2-METER CHANNELS

If you've done some operating or listening on the crowded lower-frequency bands, the first thing you'll notice about 2 meters is that the band is channelized into specific frequencies exactly 30 kilohertz apart from 146.01 to 147.99 megahertz. The first channel is 146.01, the following ones are 146.04, 146.07, 146.10, 146.13, and so on.

Confused by these numbers? They are easier to understand if they are expressed in hertz. Thus, 146.01 megahertz is the same as 146,010,000 hertz, and 30 kilohertz is the same as 30,000 hertz. Add them, and you have 146,040,000 hertz, or 146.04 MHz. The last zeros of the numbers are dropped as a matter of convenience.

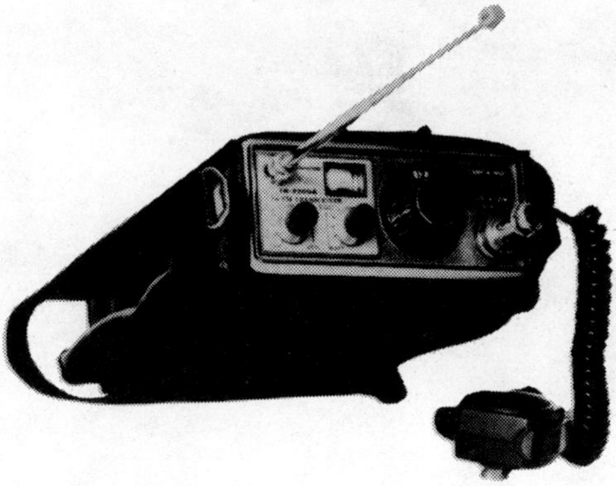

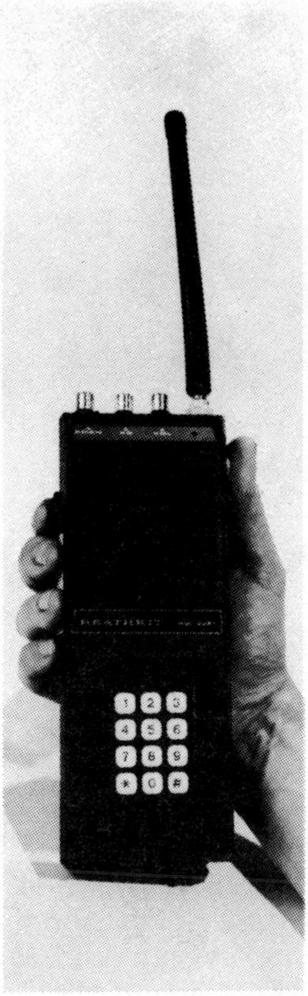

Examples of low-power 2-meter transceivers. Top: Kenwood TR-2000A slings from the neck, leaving the user's hands free. Right: Heathkit Model HW-2021 "handy-talkie" with touch-tone pad that gives access to telephone lines if the repeater station is properly programmed for this service.

2-METER EQUIPMENT

A typical solid-state 2-meter transceiver, suitable for either mobile or fixed station use, is about the size and shape of an ordinary cigar box. Its power rating is about 10 watts, and in mobile applications it needs only 2 or 3 amperes from the standard storage battery found in cars, trucks, boats, recreational vehicles, etc. This current is less than that drawn by the parking lights on most cars, so there is little danger of running down the battery in normal driving or during rest stops.

Many fm transceivers use crystals for close control of the transmit and receive frequencies; that is, each repeater pair requires two frequencies. This arrangement is satisfactory if a town or city has one or more repeaters that cover it with good signals. However, if you go on a trip, which is just when you want to operate mobile, you may find that the next available repeater operating on your home frequencies is many miles distant.

Heathkit 2-meter transceiver is within easy reach of driver of recreational vehicle *but does not obstruct his vision.*

A more sophisticated type of frequency control called *synthesizing* is rapidly replacing paired crystals because it can produce literally hundreds of frequency combinations at the twist of a knob. This is ideal for mobiling, as the author learned in 1979 when he drove the full length of Interstate 10 from Florida to Cali-

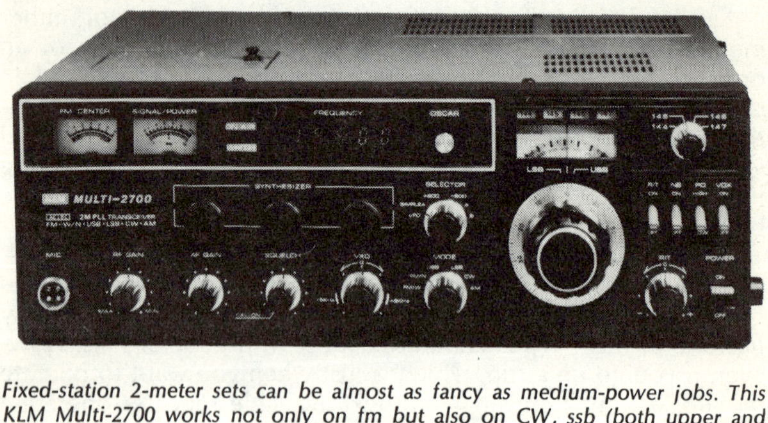

Fixed-station 2-meter sets can be almost as fancy as medium-power jobs. This KLM Multi-2700 works not only on fm but also on CW, ssb (both upper and lower), and conventional am.

fornia with a synthesized rig; at all times he was able to key a repeater and make interesting contacts.

For home use, most transceivers require an outboard ac power pack. However, some models have the power pack built in. As an alternative, you might consider the combination of a storage battery and a small trickle charger that can be connected to the transceiver more or less permanently. Practically every auto service station has a couple of used car batteries lying around for sale at reasonable prices. They might have been retired from starter use which calls for 200 or 300 amperes, but with a load of only 2 or 3 amps they could last for years.

Ham clubs put a lot of time and effort into the construction and operation of repeater stations. This association has received awards for the public service rendered by members. Exhibit was displayed at Florida convention. Note small transceiver on belt of Jim Hinchee, K4KMU.

ANTENNAS FOR 2 METERS

Since a half-wave antenna for this band is only one meter (39.37 inches) long, simple whip antennas work very well in both mobile and fixed installations. Multielement beams are easy to make and handle, as they are compact and light, but they are advantageous only for DX work, a relatively small part of 2-meter operation.

PORTABLE TRANSCEIVERS

For portable use as distinguished from mobile use, there are fm transceivers so small they can be carried in a coat pocket or hung about the neck like a camera. Their signal output is low, one watt being about the average, but their performance through repeaters is astonishing. These "handy-talkies" are of particular value for disaster work, athletic meets, picnics, road races, parades, and other outdoor events requiring close supervision. A word of caution here: Small as they are, these transceivers are still amateur radio stations, and must be handled only by licensed operators. The power source is a group of small flashlight cells, preferably of the alkaline or rechargeable NiCad types.

REPEATERS

Two-meter fm was given a terrific boost with the appearance of "repeater" stations, most of them the property of ham clubs. A repeater is a combination of a receiver and a transmitter that picks up signals from two mobiles or portables, or from fixed stations under some circumstances, and retransmits them alternately *on another frequency* at high power from a high antenna. With this arrangement, solid communication between the field stations (such as a mobile station and a fixed station) is possible over otherwise prohibitive distances.

The signal from the calling station, called the repeater *input,* and the repeater's own signal, or *output,* are always separated by 600 kilohertz. Different combinations are used in different parts of the country, but the nearest to a universal pair is 34/94; that is, 146.34 and 146.94 megahertz. Altogether, 54 channels are reserved for repeater purposes, and 13 for direct communication, or *simplex.* (Simplex means normal transceiver operation between two stations on one frequency.)

Repeaters can be either manual or automatic. In the first type, a club member is actually at the station or he can be at home with full remote control of the equipment by means of a radio link or a direct wire circuit. The usefulness of the installation obviously depends on the availability of volunteers for the job. Most repeaters are going automatic, with full 24-hour service.

Some automatic repeaters offer mobile and portable 2-meter fm operators direct access to the country's telephone system, by way of a transmitter accessory that generates the same audio tones used on land lines. The applications of this "autopatch" facility can be sensational. Imagine getting through *directly* to the police in case of a real emergency! However, incoming calls on the land

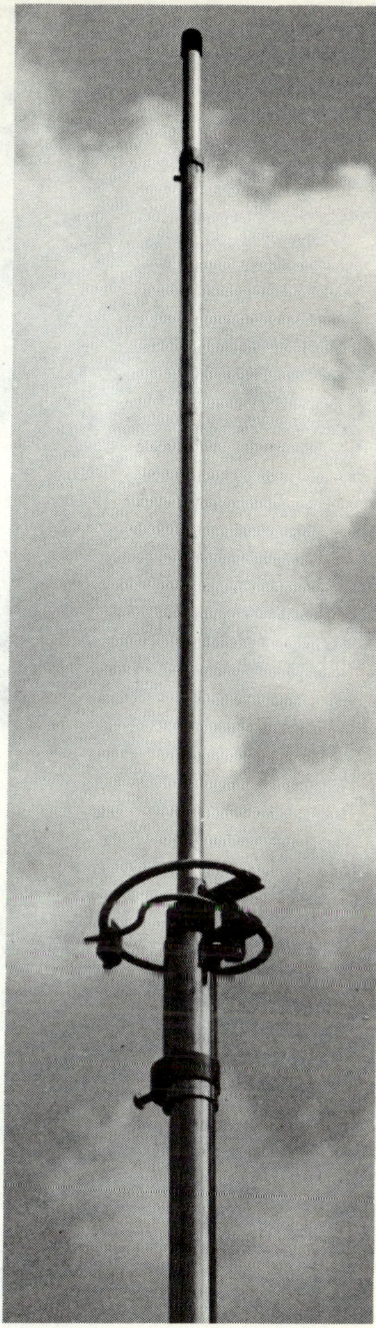

This Cushcraft "FM Ringo" antenna is widely favored because it is only 36 inches long, easily tuned by an adjustable top section, and mounts readily on tv-antenna mast sections.

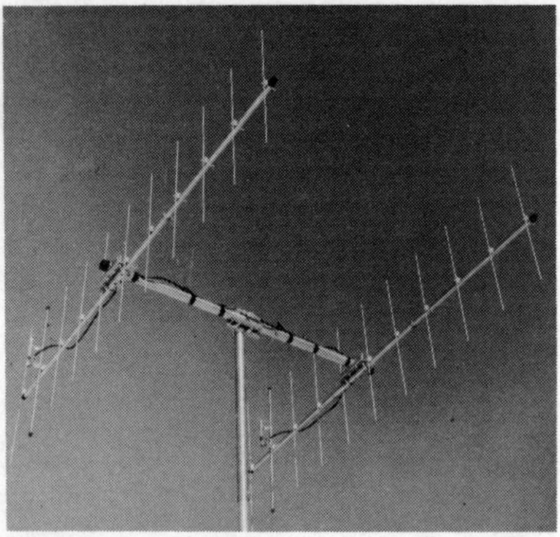

This Cushcraft "Power Pack," containing 22 elements, is a highly directional beam that appeals to experimenters who want to learn how far they can work on 2-meter fm with the minimum of power.

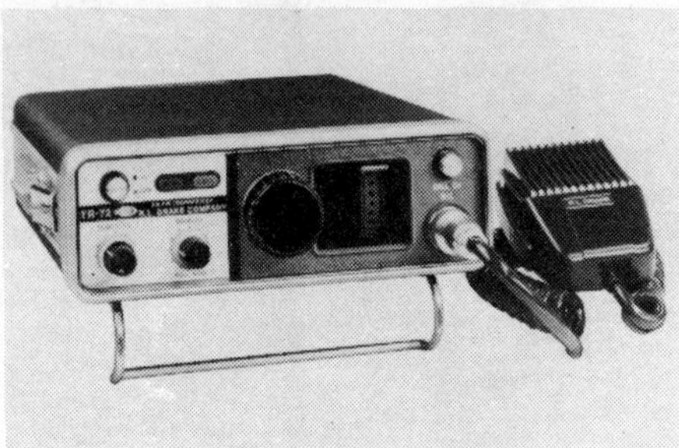

With an output of 10 watts, the Drake TR-72 is suitable for either fixed-station or mobile use.

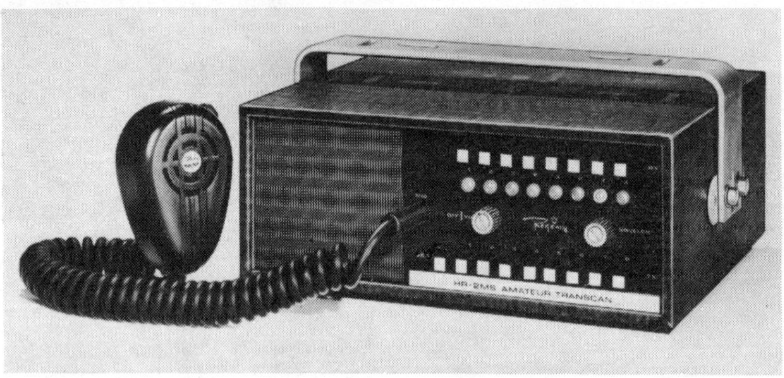

The Regency Amateur Transcan automatically scans any combination of 8 channels, locks on and listens, waits for a return signal, and resumes scan if no signal is received. Operates on 12-volts dc or 115-volts ac with optional power supply. Power output is 15 watts.

lines must be screened by an operator at the repeater; no operator, no calls from an irate wife looking for a delayed husband!

OPERATING PROCEDURE

Operating 2-meter fm through repeaters is very simple. You turn on your transceiver, punch up the frequencies for your local station, and listen. If you don't hear anything, press your microphone button once and release it. If you get through to the repeater, you will hear its carrier as a soft, steady hiss, and the meter on your transceiver will register it as a strong signal. This means that you have "keyed the machine." Squeeze the mike button again, hold it down and start talking.

"This is K3XYZ on the repeater."

No CQs, no tiresome repetition of your call sign; just this brief invitation. The channel was wide open when you spoke, so anyone who had his receiver on heard you clearly. If he's free to chat for a while, he won't come back to you right away, but will wait about five seconds for the repeater to go off. He will press his mike switch quickly to recapture the transmitter.

"K3XYZ, this is W3ABC. Handle here is Jerry and the QTH is Smithtown. Haven't heard your call before. Are you in the area? Over."

"W3ABC, this is K3XYZ. Thanks for coming back at me Jerry. My name is Harry and I live at the west end of town. Am just checking out a new rig here, an HW73. Does the audio sound OK?"

"Sure, Harry. Nice and clean. If you're new in town maybe you don't know about the Monday night swap net. Got any old equipment for sale or trade?"

Note that the call-up can be dropped after communication has been established. Also remember that you should identify at least once every ten minutes. This is a very long time to hold a repeater, so it's better to throw in both call signs every couple of minutes.

Suppose that you have to answer the door bell or the telephone. Make the next transmission snappy.

"Jerry, I have to run. We'll be in touch again. W3ABC from K3XYZ. Clear of the repeater and pulling switches."

The last three words mean that you are shutting off your transceiver. If you and Jerry merely run out of conversation but you intend to continue listening after signing, omit "and pulling switches."

The main purpose of replying after a brief delay is to permit mobile operators to break in with emergency calls, invariably concerning road incidents. If you hear an excited voice shouting "Break, Break," you'll know that something serious is afoot.

After you've been talking back and forth for a while, don't be surprised if you suddenly find yourself off the air. Most repeaters have automatic timers that limit conversations to a minute or several minutes, depending on the ham population likely to be using the facilities. This interruption is a pointed way of suggesting that you should pipe down and give others a chance to carry on QSOs. If no one takes advantage of it, you are free to key the machine again, of course.

During a QSO, you are quite likely to hear a recorded identification of the repeater call sign, either in CW or voice. This lasts only seconds and does not affect the live signals from you and the station you are working.

DX ON 2 METERS

Transmission on 2 meters is generally limited to line-of-sight distances. However, hams by nature are DX hounds, so many of them build up elaborate beam antennas, mount them on high towers with rotators, and spend a lot of time and energy trying to reach other 2-meter DXers over the horizon. They are sometimes aided by atmospheric conditions, and sometimes blocked by them. It can be unnerving, but also very interesting, to hear a station 200 miles away access your local repeater. It might fade out in the middle of a word, then return slowly or suddenly.

There is much DX being worked on simplex; that is, direct station-to-station on 146.52 MHz without going through any repeater. Conventional operating procedure is used, with no time limit.

MOUNTING THE TRANSCEIVER

Most low-power 2-meter transceivers that are intended primarily for use in cars are sold with U-shaped brackets to permit hanging them from below the dashboard. However, this seemingly simple job, which involves only a couple of nuts and bolts, can be difficult, impossible, or inadvisable in some vehicles because of the way they are built. The underside of the dashboard is often a sort of closed box containing various wires and cables and, if you drill through it blindly, you might cut into the latter. Even if you don't, you might not be able to reach the bolts to get the nuts on them. Toggle bolts might be one alternative, but with

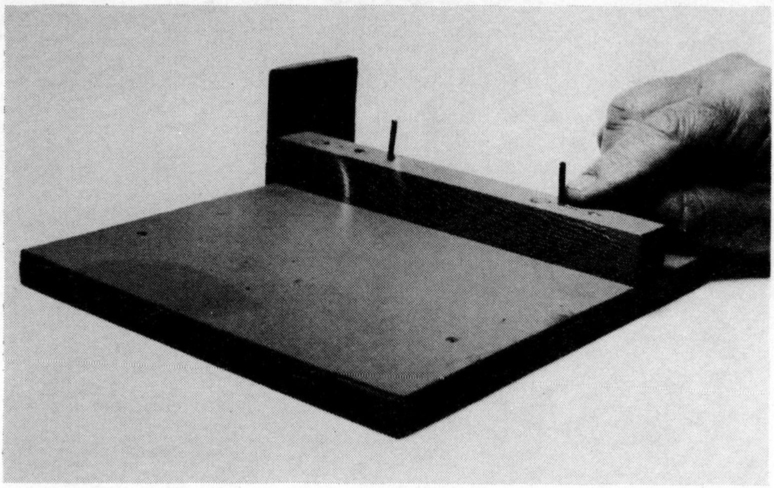

This board is half a flat mount for a 2-meter transceiver. The finger points to one of two headless nails driven part way into a one-inch strip fastened to the board. These nails will mate with holes in the front edge of the other half of the mount. The small square piece at the left is a stop against which the latter fits.

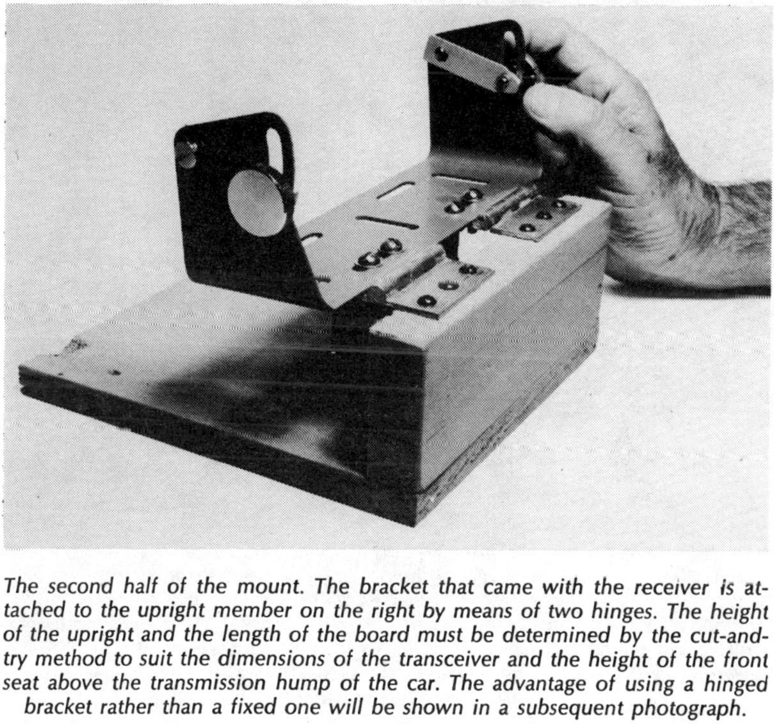

The second half of the mount. The bracket that came with the receiver is attached to the upright member on the right by means of two hinges. The height of the upright and the length of the board must be determined by the cut-and-try method to suit the dimensions of the transceiver and the height of the front seat above the transmission hump of the car. The advantage of using a hinged bracket rather than a fixed one will be shown in a subsequent photograph.

Here the transceiver is secured in its bracket with its rear end just clearing the back edge of the L-shaped base. Space must be allowed here for the antenna and battery connections. Note how the right end of the base fits neatly over the nails in the other section. Depending on the area of the transmission hump in a particular car, it might be possible to use a single board for the base. In most cars, however, the two-piece construction is more flexible.

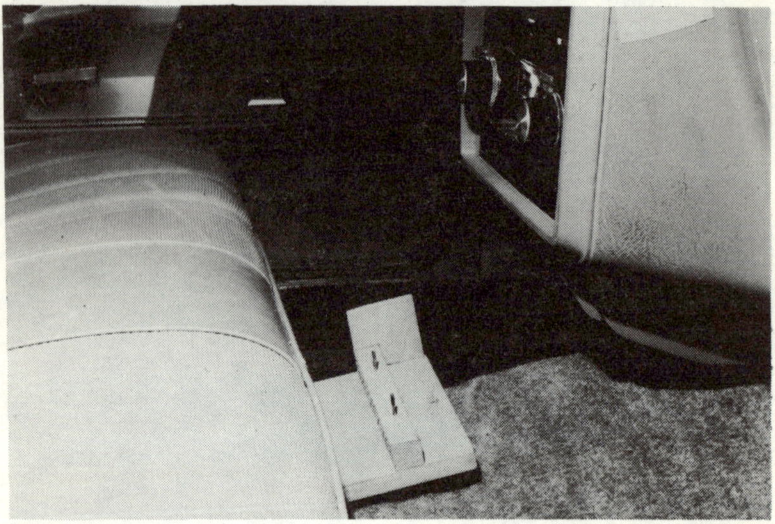

Here is the start of an actual installation is a typical car, a Chevrolet sedan, which has a rather large console over the hump containing a broadcast receiver, heat and air-conditioning controls, an ash tray, and air ducts. The "tongue" of the mounting base is about to be pushed under the seat with the small horizontal strip against the seat's front surface. To assure a tight fit, shims of thin plywood or strong cardboard can be added to the board, which is then left in position more or less permanently.

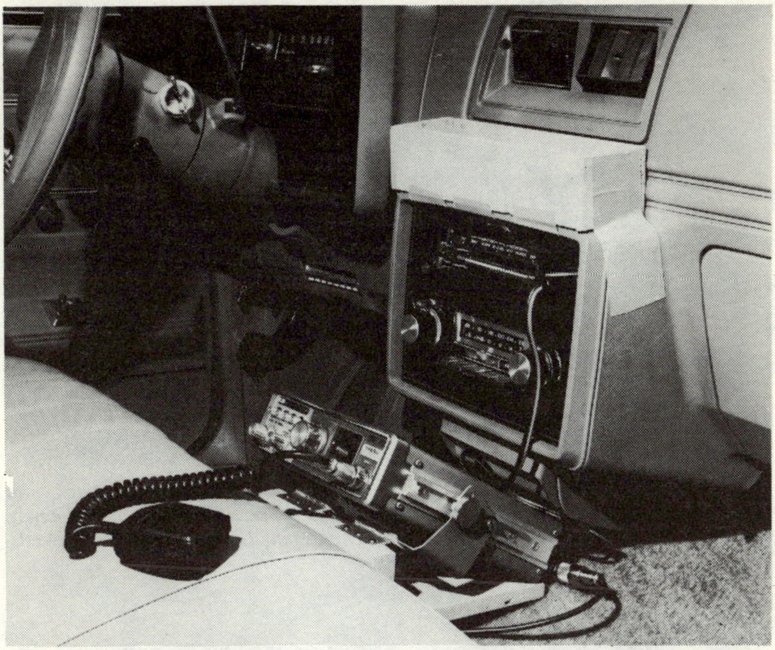

Here the transceiver is ready to go on the air. Note how snugly it fits between the seat and the ash tray at the bottom of the console. Also, note how open and accessible the controls and microphone are. The 12-volt cord plugs into the cigarette-lighter receptacle on the console, and the coax cable from the antenna plugs into the right-angle coax adapter on the back of the receiver. With the tongue left in place, installing or removing the other section takes less than ten seconds. With this neat layout, 2-meter fm is a real pleasure!

them the possibility of damage is even greater because the wing nuts cover a larger area than ordinary nuts.

From the operating standpoint, the best position for the transceiver is the center of the dash, because here it is within easy reach of both the driver and the front-seat passenger. There is really no need to drill holes in the car. As shown in the accompanying illustrations, the idea is to make a simple box or cradle for the transceiver out of scraps of wood and to sit the container on the transmission hump. In either case, the transceiver itself is angled so that the controls are comfortably in sight.

With the hinge-mounted transceiver removed from the car, the back end propped up with a small block of wood, and a dc power supply placed on it, the result is a fine home station that will provide many hours of pleasant contacts.

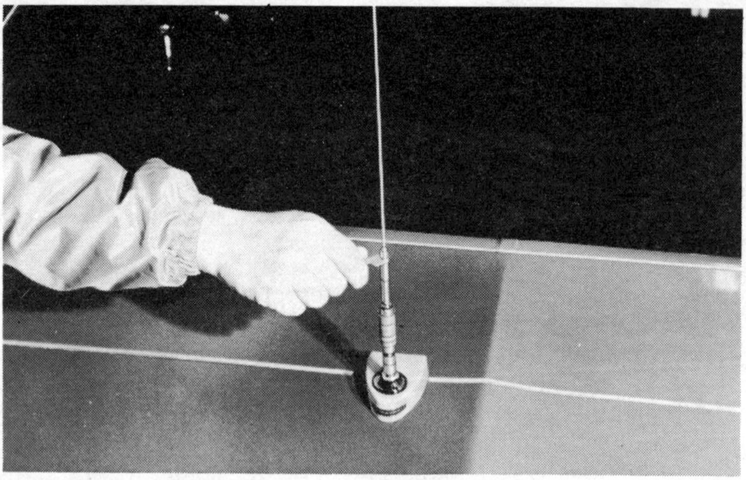

Mounting a whip antenna for 2-meter fm is quick and simple with a base that clamps to the back edge of the trunk lid. The coax cable passes through the car body to the transceiver up front.

Luggage-rack bar makes an excellent support for one or more antennas. The slender whip in the center is 53 inches long and puts out a potent signal on 2-meter fm because it is high and clear. A "grounding" wire from the antenna base is stapled to the underside of the bar and runs to the support brackets clamped to the rain gutter. Note the additional antenna base at the right.

ANTENNAS FOR MOBILE OPERATION

Probably the most popular antenna for 2-meter mobiling is the ⅝ *wavelength* whip. This is about 53 inches (1.35 meters) long, has a ⅜ × 24 threaded end, and is usually mounted on the hinged lip of the trunk lid by a U-shaped clamp. It assumes a marked backward angle, depending on the contour of the car. While this position is convenient, it is not always the most effective. As experimentation with a small field-strength meter will quickly show, radiation from the antenna can be noticeably better in some directions than in others. This is not desirable in mobile operation, because the relative positions of two stations in contact keep changing.

With most cars, a more uniform radiation pattern and generally improved performance are obtained if the whip is mounted in the center of the car roof. Here the antenna is unobstructed and appreciably higher than before; in fact, it is close to being the ideal antenna. The problem is how to anchor it without having to make a hole in the roof. Rubber suction cups have been used, but with limited success. Magnetic mounts hold well if the bottom surface of the magnet is perfectly flush against the top and if the latter is *not* vinyl covered.

Is that a CB antenna? It certainly is. Many hams also own CB rigs, and there's no reason to leave them at home if a place can be found for them in the family car.

The second antenna base is very convenient for experimenting with antennas of various lengths and shapes and with various transceivers. This little whip, made from a coat hanger, works very well on 2 meters.

A mobile tuning meter is invaluable for 2 meters because it provides a positive indication of overall antenna and equipment performance. This representative model contains only a few small components and doesn't even require a battery. It's like an old-fashioned crystal receiver that brings in several broadcast stations at a time. The antenna is a piece of stiff wire of any length between a foot and three feet, and it plugs in at the top of the case.

An arrangement that is inexpensive and also great for experimental purposes uses a single wooden or metal bar of the type usually sold in pairs for luggage racks. The wooden bar shown in the accompanying photographs cost all of one dollar at an auto junk yard. It is held securely by end clamps that grip the rain gutter. The bar can be loosened and moved back and forth to determine, with the help of the field-strength meter, the position that gives the strongest radiation.

A mounting base previously used on the trunk lid is screwed to a small aluminum plate cut from a discarded radio panel, and this in turn is fastened to the bar with wood screws. The coax cable from the base is stapled or taped to the underside of the bar and enters the car through a slit cut in the frame gasket of a window. With the ⅝ whip installed on this elevated mounting, the owner of the rig could "access" repeaters he couldn't even hear before.

A 53-inch (1.35-meter) whip on the top of a conventional passenger car can be annoying when it encounters low tree branches, gas station canopies, and the like. It can be replaced with little loss, for local communication via repeaters, by a ¼ wave antenna. For the 2-meter band this antenna is only about 19 inches (.5 meter) long. You can make one in minutes by straightening out a wire coat hanger, cutting it to about 21 inches (53 centimeters),

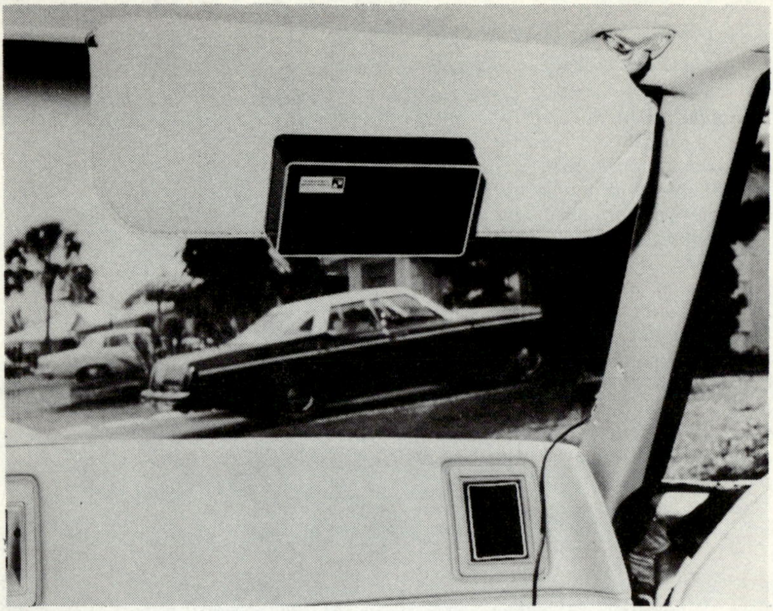

When the windows of a car are open and traffic noise is heavy, mobile operation can be a bit difficult. It can be much more tolerable if a clamp-on visor loudspeaker is used instead of the one in the transceiver. Most sets already include a speaker jack on the back panel, so connection of the outboard unit is simple.

and attaching one end to an alligator test clip. Set the alligator clip in the antenna mount, and place the field-strength meter on a chair a few feet from the car, in the open. Fire up the transceiver, press the microphone switch, and note the meter reading. Release the mike switch, snip ½ inch (1.25 centimeters) from the wire, press the mike switch, and again note the reading. If it's a little higher than before you are on the right track, so repeat the procedure ½ inch (1.25 centimeters) at a time. The reading will keep rising, and then start to fall. The antenna will now be too short, so make another one and cut it to the length that gave the maximum meter indication.

13

Electronics as a Career

Many young men take up ham radio as a hobby and become so engrossed in it that they decide to make electronics their life work. This is a smart move, because the electronics industry in its multiple ramifications offers almost unlimited opportunities to serious, diligent workers.

HOBBY INTO BUSINESS

Former and present hams are scattered throughout every level of the radio business: management, engineering, manufacturing, merchandising, advertising, television and sound broadcasting, and servicing. The history of the art is full of the names of hams who made good in a big way. Major Edwin H. Armstrong, whose invention of the regenerative, superregenerative, and superheterodyne circuits and of fm broadcasting easily rates him the title as "Radio's No. 1 Wizard," always spoke proudly of his pioneer amateur station of 1906 in which he started his famous research. Allen B. DuMont, whose name is synonomous with television and the development of the cathode-ray tube, was an active ham in the late 1920s while he was an engineer with various radio firms. The huge organization that he started was the direct outgrowth of a modest laboratory-shop he had in the basement of his home in Upper Montclair, N.J., in 1931. Sound broadcasting as we know it today had its beginnings in 1920 with 8XK, a ham station in the backyard garage of Frank Conrad in Wilkinsburg, Pa. This modest "shack" grew into KDKA—the station that made the United States aware of the new miracle of radio.

Conrad eventually became assistant chief engineer of the Westinghouse Company, and at his death was mourned as the "father of broadcasting." The great Marconi himself, the man who literally electrified the world at the turn of the century by sending a radio signal across the Atlantic, was scarcely 20 years old when he undertook his "wireless" experiments with odds and ends of equipment, just as many an aspiring ham does today!

HAM INTO ENGINEER

How does a ham become something more than a ham, careerwise? For a young man—or young woman—whose objective is engineering, the way is very clear: After graduation from high school, a sojourn of four years at a recognized college offering a bachelor of science degree, preferably in electrical engineering. The various courses offered by colleges throughout the country are listed in their catalogs, which are obtainable upon request;

These three men are hams, but they are also highly trained field engineers who keep Raytheon's complicated radar systems around the world in working order. The shack belongs to "Pops" Karentz, W1YBL, center. At left is George Lucas, W1ZYS; at the right, Ray Churchill, W1BVI.

they are also available in most local high schools or public libraries. The catalogs show entrance requirements, tuition fees, etc.

For many years, college instruction in "electrical engineering" was devoted largely to power engineering, with little attention given to anything beyond 60-hertz ac. Today, however, a student can elect to take a great deal of communications and electronics in his junior and senior years. After graduation with a basic B.S. degree, he can take advanced courses in electronics at universities having modern laboratories and other facilities. If his money, his eyesight, and his girl friend don't disappear along the way; he can work up to the degree of master of science (M.S.) or doctor of science (D.Sc.) and earn the right to be addressed as "Doctor Whoozis" in academic and engineering circles.

Oddly enough, in spite of the size of the electronics industry and the great number of engineers occupied in it, there is no such thing as a degree in either "radio engineering" or "electronic engineering."

ENGINEERS VS. TECHNICIANS

By commonly accepted standards, an *engineer* is a college-trained B.S. (or better) who can undertake original design or research on components, materials, circuits, and complete pieces of equipment. While a degree is no guarantee of anything other than the fact that its holder got through school, it is the primary basis for selection of engineering employees by all radio-electronics firms.

There are thousands of men who are graduates of good vocational high schools, of military electronics schools, of private commercial schools specializing in electronics, and of the hard school of experience. Many of them know more about the practical side of radio than the holders of three degrees. Regardless of their background, they are not engineers; instead, they are classified as technicians. This is a broad term, but in the eyes of executives it distinguishes these men from the college group. For example, a help-wanted advertisement in the Proceedings of the Institute of Radio Engineers specifically reads: "Engineers—E.E. graduate with 3 years experience. Technicians—2 years technical school in communications and 3 years experience."

Technicians are in great demand for responsible jobs in factories, laboratories, sales organizations, radio and tv stations, service shops, etc. Some men are so competent that they are often called "engineers without degrees." However, make no mistake about it: on the pay, recognition, and social levels, technicians definitely rate below full-fledged engineers.

Many skilled technicians are entirely self-taught. A man can start as a ham at the age of 15 or 16. By studying the excellent "amateur handbooks" available for a few dollars and accumulating intensely valuable experience with receivers, transmitters, test equipment, etc., he often is a real "expert" by the time he reaches voting age. He probably has more savvy at this point than a college student of the same age just getting his degree. However, his lack of formal "book larnin" shows up painfully when the going gets really rough. For example, he may be able to assemble, install, and repair a very complicated piece of equipment like a radar set, but he's lost if he tries to interpret certain readings. Radar is a matter of angles in space, and a man must know advanced trigonometry pretty thoroughly to solve the problems. Here is where an engineer steps in and earns his keep. He is able to solve it because he has had trig and a dozen other forms of difficult math drummed into his head all during his college work. It is generally conceded that mathematics is the dividing line between the true engineer and the technician.

Some personnel executives are of the opinion that the emphasis on math in many colleges is too great, and at the expense of practical training in the laboratory, shop, or field. This tendency is probably the result of industry's demand for brainy characters who can use the higher forms of math to solve the fantastically complex questions that arise in connection with guided missiles, radar, telemetering, computers, color television, etc. This work is so advanced that engineers who took their degrees ten or fifteen years ago find themselves unable to cope with it. To keep their jobs, they take refresher courses at night, sometimes quietly and without telling anyone about it.

TECHNICIAN TRAINING

As a ham, you have picked up a fair working knowledge of equipment; now you want to advance into the professional side of radio. Suppose you can't afford to go to college, or feel that you are too old to start. Is there still a worthwhile career for you ahead? Most definitely, yes. There are excellent resident schools in several sections of the country where you can get concentrated, short-term instruction in one or several phases of electronics, up to almost the full engineering level. Upon completion of the course or courses, you are fully qualified to work as a service technician, factory-test man, field-installation man, radio-station attendant, etc. If you like to get around and have had any previous experience in selling, you will find good opportunities in the technical sales field. If foreign travel appeals to you, you'll be snapped up in an instant for installation, maintenance, and instruction work on American-made electronic equipment sold for both military and civilian purposes to numerous countries.

If you can't get away from home to attend an out-of-town school, you can do very well with correspondence courses. Many of the independent service technicians now in business got their training by mail, and good training it is. The lessons are carefully organized into small units, and you can go through them as slowly or as quickly as your learning ability permits. With most courses you receive actual equipment that you assemble and wire and then use for actual testing and measuring applications.

One very important feature of correspondence training is financial. You can pay for most courses as you go along, instead of having to dig into the family savings for a big chunk of money all at once. Many students of servicing courses pay them off with money they earn applying the lessons almost as soon as they complete them.

Even the correspondence courses can be pursued to a high level. You can take so much math and advanced theory that you finish up as a "practical engineer."

CAREER SWITCHING

Some men who take up ham radio as a hobby switch to electronics as a business even though they are well along in years and have been earning their living in completely unrelated activities. They usually prepare themselves for their new careers by taking correspondence courses or attending night school. The reason for these changeovers is simply the excitement, glamour, and prospects that the electronics industry offers. In spite of the startling developments in television, radar, two-way communications, etc., in recent years, the immediate future promises even more. The opportunities should prove excellent for those who have a good background in electronics.

Highly successful shifts have been made by the most unlikely people. Files of most schools bulge with testimonials from ham students who have made good and are happier in their new vocations than they ever were in their old. The author knows personally of a stock broker who became service head of a large radio manufacturer; a furrier who now runs a prosperous high-fidelity sound business; a clothing salesman who is now purchasing agent for a radio distributor; a lawyer who is now a high-ranking electronics officer in the Navy; a retired brigadier general of the Regular Army who runs an appliance store; and a policeman who now specializes in auto-radio installations.

HAMS WHO HAVE MADE GOOD

Wherever you go in the electronic field, you are sure to find hams. An astonishing number of heads of firms maintain their calls and are active on the air.

For instance, there is Albert Kahn, K4FW, founder of Electro-Voice, Inc., for many years an outstanding manufacturer of microphones, speakers, and various high-fidelity sound components. Al is now president of Ten-Tec, Inc., which specializes in amateur radio gear including inexpensive ham equipment for beginners. You'd think he gets enough radio during his normal working day, but that doesn't keep him from enjoying a hobby he has pursued since 1918. He runs a full kilowatt station down in Sevierville, Tenn., and has a well-deserved reputation as a fast man with a telegraph key.

Albert Kahn, K4FW, president of Ten-Tec, Inc. at the controls of his ham station. With his right hand he is operating an electronic keyer which makes dots and dashes semiautomatically.

Another "brass pounder" is Russell McFall, W3JAB. Appropriately enough, he is president of the Western Union Telegraph Company!

Another ham who has made good is William J. Halligan, founder of the Hallicrafters Company of Chicago. He was an active ham at the age of 14, and a licensed commercial operator

William J. Halligam, W9AC, the founder of Hallicrafters Co.

at 16. He served as a brass pounder in the Navy in World War I, attended West Point for three years, and resigned to get married. He was a radio writer for a Boston newspaper and then a sales representative for various radio firms in Chicago.

During the depths of the depression in 1933, he started Hallicrafters with the object of providing hams with good receivers and transmitters at reasonable prices. The company had made a fine reputation for itself when World War II came along and skyrocketed its business to heights never dreamed possible. The production of high-grade amateur equipment was resumed at the end of the war.

Bill has retired from the company, and divides his time between Illinois and Florida. He is still active on the air as W9AC.

On the engineering, laboratory, installation, and service levels there are more hams than you can count. Some men who let their licenses lapse years ago are still very much hams in spirit, and do a considerable amount of short-wave listening even if they are prevented by circumstances from operating transmitters.

ELECTRONICS SCHOOLS

A complete listing of schools other than standard colleges and universities that offer specialized training in electronics in its various phases can be obtained from either *National Association of Trade and Technical Schools,* or *National Home Study Council,* 2021 L Street N.W., Washington, D.C., 20036. A letter addressed to either institution will assure you of obtaining the latest up-to-date catalog of resident or correspondence schools.

14

The Radio Marketplace

Ham equipment is not sold by the same dealers who handle television and radio sets, but by merchants who are known in the trade as "electronic parts distributors," or "jobbers." Most of them carry large stocks of electronic components of all kinds, which are purchased by service technicians, students, experimenters, schools, and industrial firms, as well as amateurs. In some large population centers, there are dealers who specialize exclusively in ham gear.

To locate a source of supply near you, the first thing to do is consult the Yellow Pages of the local telephone directory. Look under "Radio Supplies and Parts"; this listing may be followed by the notation "Wholesale & Manufacturers" or "Retail Outlets." Also, look under "Radio." By making a couple of phone calls, you'll quickly find a store anxious to serve you. Practically all amateur equipment suppliers employ hams on their sales staff, and these men can be of great assistance to a newcomer.

Some of the larger radio distributors issue annual catalogs. Many of these run to several hundred pages, are veritable encyclopedias, and are well worth having. Check the advertising columns of any of the ham and electronics magazines for the names and addresses of firms that offer free catalogs.

Shortly after you qualify for your ham ticket, you will probably start to receive catalogs and literature by the boxes full. The names and addresses of new licensees are a matter of public record, and are used by enterprising firms to obtain new customers.

HOW TO BUY

Except when the catalog specifically states "prepaid," published prices do not include the cost of shipping merchandise directly to the customer. Hams sometimes misunderstand this practice, but it is standard in the field. Obviously, if a jobber's store is within convenient distance, you can save money by shopping in person. Many jobbers stay open one or more evenings a week to accommodate buyers who can't get away from their jobs during the day.

Small packages are sent most economically by UPS or parcel post, while large items must go by motor freight or express. Because of his experience in these matters, let the jobber recommend the method of shipment best suited for your merchandise and for your location in relation to his warehouse. There may be some charges for insurance and Federal taxes on shipping, but these are usually only minor.

TIME PAYMENTS

After reading the descriptive copy in catalogs, you may readily take a shine to some very fancy—and also very expensive—ham gear. The fact that you do not have all the ready cash may not lessen your infatuation. It might be welcome news to you, therefore, that practically all jobbers sell merchandise on credit. Banks and finance companies handle the transactions just as they do in the case of cars, houses, boats, etc.

USED EQUIPMENT

There is a vast turnover of equipment among hams as they graduate from one type of station to another or as their interests change. Many jobbers have excellent bargains in "trade-ins." The classified advertising columns of the ham magazines are also full of tempting offers to sell or swap.

As a hobby, ham radio seems to be closely related to shooting and photography. It is interesting to read the classified ads and to note that receivers, transmitters, rifles, pistols, cameras, and enlargers are always for sale or wanted.